Mobile Radio Technology

Mobile Radio Technology

Gordon White, MCGI, CEng, MIEE

B H NEWNES

Newnes
An imprint of Butterworth-Heinemann Ltd
Linacre House, Jordan Hill, Oxford OX2 8DP

℞ A member of the Reed Elsevier group

OXFORD   LONDON   BOSTON
MUNICH   NEW DELHI   SINGAPORE   SYDNEY
TOKYO   TORONTO   WELLINGTON

First published 1994

British Library Cataloguing in Publication Data
White, Gordon
   Mobile Radio Technology
   I. Title
   621.3845

ISBN 0 7506 0931 1

Library of Congress Cataloguing in Publication Data
White, Gordon.
   Mobile radio technology/Gordon White.
   p. cm.
   Includes index.
   ISBN 0 7506 0931 1
   1. Mobile communication systems. I. Title.
   TK6570.M6W49   93–33162
   621.3845–dc20   CIP

Typeset in Bembo by �𝕿 Tek-Art, Croydon, Surrey
Printed and bound in Great Britain

Contents

Preface

Our way of life since the Industrial Revolution has been determined not only by the laws of the country we live in but by a continuous list of inventions and discoveries. Throughout most of history the world appeared the same when one was born as when one died. In the last hundred years, unless one has lived far from recognised civilisation, the world has appeared to change many times throughout one's life. Two world wars and confrontation between two major powers for the last 50 years has given the impetus for considerable technological developments which have then found their way into our everyday life. It is difficult to visualise the way of life in the last century without cars, aeroplanes, radio, television, telephones and even electricity. These inventions now control the lives of people in the more developed countries and every underdeveloped country strives to emulate the technically advanced countries.

Communications have shrunk the world so that countries once remote are at the most only 24 hours away by aeroplane or seconds away for telephone conversations, radio or television. Unless there is a major world catastrophe the poorer nations must one day reach the same standards that are enjoyed in the developed countries as technology must eventually spread throughout the world even if it is only to exploit the markets which the vast populations of the technically underdeveloped countries represent.

Apart from the discoveries and inventions associated with electricity generation, it is probably the discovery of radio waves and the inventions which followed that have been responsible for the major changes in the world. Without communications many of the changes would not have happened.

Wired communication systems are satisfactory and economic where populations are dense and distances are relatively short, and for the majority of users a cable provides the connection to the communication system. Coaxial cable and more recently fibre optics have expanded these systems considerably and provide both national and intercontinental communication with cables under the sea. For long distance communication, the telephone and data links are often terrestrial microwave radio links or satellites, and our national and international systems have become a combination of all the technologies. The economics of long-distance communication links becomes acceptable with the ability to multiplex signals together so that a single cable or radio link is able to carry hundreds and even thousands of telephone conversations simultaneously.

Most people receive their radio and television broadcasting services directly from either terrestrial broadcast transmitters or satellites, although cable systems have expanded considerably in many countries and are now able to bring the viewer television from many sources on a single cable. In time most people living in a city will have the opportunity to receive cable television and to receive all the available programmes without the need for a large array of aerials and dishes. With the programmes on the cable will come the possibility of many other services such as security, meter reading, shopping, banking and interactive learning programmes.

Satellites have revolutionised communications in a major way and are now used for broadcasting, communications, navigation, spying, weather forecasting, thermatic mapping, astronomy and surveying. They have become essential for these industries and have allowed systems to cover areas which would be impossible by any other method.

Where mobile communication is required there is no alternative to radio. Our airlines and ships could not operate safely without it. The police and other public services could not be efficient without radio communication. Business has quickly adopted the mobile telephone systems and is looking for international systems for both voice and data. For the general public the personal mobile telephone may become as common as the pocket calculator.

The computer has changed the business world and the way we all work or it determines the services we receive. It has also produced the digital technology and control which is changing our conventional analogue communication networks. Digital communication provides many advantages when transmitting and processing information and there is an improved quality of signal. Digital networks can also offer many services not available on an analogue system.

Large systems take many years to develop and implement and when on a national or international scale there are many political and commercial interests, besides the technical problems, which must be solved before the systems become a reality. However, standards can be set in two ways. They can be set by committees spending many years agreeing technical specifications or a large commercial company can make a breakthrough both technically and in the market-place and everyone has to join the band-wagon or go under. We have seen this latter situation with video recorders, computers, discs and cassettes in the last 20 years. If the public likes what is available and wants it immediately it will buy it. One can never say with certainty what the situation will be in 10 years' time.

Mobile Radio Technology has been written to explain today's systems and indicate possible future developments. It aims to give an explanation suitable for students and technicians to understand the theory and practice of the mobile radio industry. It should also be of benefit to managers responsible for systems. It does not describe equipment circuitry in detail, as this will vary with the different manufacturers, but it concentrates on the principles involved. Most of the circuitry of modern communication equipment is dedicated integrated circuits and the systems are controlled by computer software. The opening chapters will explain the important basic principles of radio. These do not change, and to understand the function, advantages and limitations of any system one must understand these principles. The theory is later expanded where necessary in order to explain the individual systems.

Mobile radio is one of the great successes of recent years and there is a constant need for technicians to enter the industry as it expands. Systems are becoming international and the plans the industry has for the future will ensure the technician becomes involved with all the latest technology.

Acknowledgements

I wish to thank the many companies and individuals who assisted me in producing this book. These include:

Aerial Group
Air Call
Allgon
British Telecom
Cellnet
Digital Mobile Communications
Federation of Communication
   Services
Fleet Mobile Communications
Hutchison Telecommunications
Maxon
Mercury Communications

Motorola
MTN Magazine
Multitone
National Band 3
Paging Systems
Paging Trade Association
Philips Radiocommunications
   Agency
Southampton Institute of Higher
   Education
Vodafone

I also wish to acknowledge the technical advice given to me in discussions with many engineers and amongst these I wish to thank especially: Derrick Baker, Peter Ramsdale, Peter Richie, Gerrard MacNamee, Dave Roberts, Peter Jordan, Allen Woodham, Ray Wilmot, Robert Head, Don Baker, Philip Johnstone, Roger Lownsborough, Michael Jackson and those who have lectured at meetings or provided information at exhibitions.

The format and contents of the book are, however, my responsibility.

Gordon White, MCGI, CEng, MIEE

1 History

Geniuses often come in the form of mathematicians. Their understanding that the laws of physics that govern our world and universe can be explained with formulae and calculations has led to many predictions that have proved accurate when technology has advanced sufficiently to prove or discover the accuracy of the mathematician's supposition. Today many physicists are engaged in research on such work as atomic physics and astronomy, predicting and proving theories which originate with mathematics.

Such a man was James Clerk Maxwell (1831-1879). He was considered the leading theoretical physicist of the nineteenth century and amongst his many great works he was responsible for the electromagnetic theory of light which was to lead to the discovery of radio waves and the invention of radio.

It was 23 years after Maxwell had produced his theory that Heinrich Rudolf Hertz (1857–1894) was able to produce electromagnetic waves by electrical oscillations and prove they could travel through space at the speed of light. His transmitter was a spark coil with large plates connected to each side of the spark gap. His receiver was a wire loop with a small gap in which a spark was produced when the transmission took place. Hertz proved that the radiation behaved in the same way as light and could be reflected, refracted and polarised.

Guglielmo Marconi 1874–1937

It requires a brilliant and dedicated engineer to take theory and make it of practical use. This was the achievement of Marconi who in 1894 at the age of 20 had begun experimenting at his home in Italy. His transmitter consisted of a spark gap between two metal spheres which were attached to two metal plates. One plate was connected to earth and the other was raised high. With this arrangement he achieved distances of 1 km. His receiver consisted of the same aerial arrangement and this was connected to an evacuated glass tube with an electrode at either end and fine metal filings between. This was known as a coherer. When the signal was received the filings stuck together (cohered) and the resistance dropped. This allowed a relay to operate a tapper and inking machine which made permanent marks when the Morse code was received. In 1896 Marconi took out his first wireless patent. He then came to England and demonstrated his system to the War Office and Post Office on Salisbury Plain over a distance of 2.8 km (1.75 miles). Development continued and in 1897 he sent signals for 13 km (approximately 8 miles) across the Bristol Channel and then for twice the distance between Alum Bay in the Isle of Wight and Bournemouth. The East Goodwin lightship was the first to make use of radio telegraphy when in 1899 it used its demonstration equipment to send distress signals after it was damaged in a storm. In the same year the lightship was involved in a collision with another ship and again used the transmitter to send a distress signal. Marconi continually increased the range of his equipment and in 1899 transmitted across the English Channel between Wimereux near Boulogne and his headquarters at Chelmsford. This was a distance of over 130 km (80 miles). In 1900 the Marconi International Marine Communication Company was formed to provide radio communication to and between ships at a time when the British Navy was the largest in the world.

The spark-gap transmitter sent out signals over a wide spectrum which caused interference with other transmitters and would have prevented a large number of users. In 1900 Marconi took out a patent which enabled specific frequencies to be transmitted (Syntonic system). This proved very successful and made it difficult for others to develop systems without infringing Marconi's patent. In 1901 Marconi decided to attempt the transmission of radio across the Atlantic. Poldhu in Cornwall and Cape Cod in Massachusetts were selected as sites. At Poldhu 200-foot masts were erected but on 17 September 1901 a severe gale destroyed the aerials. Within eight days new aerials had been constructed and tested. The transmitter operating at 20 kW DC input power was more powerful than any transmitter previously built. (The DC to RF conversion rate, it is estimated, may have been about 20% for a spark transmitter.) With commercial pressures building up on Marconi, he abandoned a two way transmission test and moved the reception site to St Johns in Newfoundland after the aerials were also blown down in Cape Cod. While the Poldhu transmitter continued to send the letter S in Morse code (three dots) Marconi and his assistant flew their aerial on a kite at the receiving site in more gales. It was on 12 December that Marconi finally heard the weak signal amongst the high static noise on a telephone head set but it was too weak to operate an inking machine and Marconi had no physical proof. However, his achievement was recognised by the American Institute of Electrical Engineers and in the following year Marconi equipped the *SS Philadelphia* and sailed to America. While sailing he received signals from Poldhu at a distance of over 2000 miles and proved the strength of the received signals increased at night and also followed the curvature of the earth.

For his work Marconi shared the Nobel Prize for physics in 1909. In 1929 he was made an Italian senator. A memorial now stands on the cliffs at Poldhu from where Marconi made his transatlantic transmission.

The Russians also claim the discovery of radio as Aleksandr Stepanovich Popov, a physicist who lived from 1859 to 1905, succeeded in transmitting radio waves in 1897 over a distance of 5 km. He did not, however, pursue the use of them for communications but used them to study thunderstorms.

The growth of telecommunications

The story of telecommunications follows four different paths which converge and diverge with different applications. The principal technologies are telegraphy, telephony, radio and data. Each has made progress as invention, materials, manufacturing technology and the requirement to communicate all occurred together. Necessity is known as the mother of invention and it is not surprising that rapid progress is made during times of war and those periods of the Cold War when money for development was readily available. Technology applied to civilian uses is often the spin-off from military development or prestige projects such as putting a man on the moon or Star Wars.

Communication history using electrical signals began in 1800 with the invention of the battery by Volta and the discovery by Oersted that an electrical current flowing in a wire could deflect a compass needle. This became the basis for the early telegraph systems. A telegraph system had been demonstrated as early as 1774 by George Lasage of Geneva who used an electrostatic machine and one wire for each letter. In 1816 a superior single wire electrostatic telegraph machine was demonstrated by Sir Francis Ronalds in London but there was no interest when he offered it to the Admiralty. The invention of the telegraph coincided with the growth of the railways. In 1837 Cooke and Wheatstone obtained a patent and later demonstrated a five-needle telegraph whereby letters were arranged on a board in pyramid fashion and two needles were operated simultaneously to point to a particular letter. In 1839 this was installed between Paddington and West Drayton by the Great Western Railway (GWR) and later extended to Slough in 1842. The public was also invited to use the system. In 1846 Cooke and Wheatstone simplified their telegraph to a single needle which required only one wire. In the USA Professor Morse of Art and Design at the New University of New York devised a telegraph which recorded messages on paper tape, and one of his students, Alfred Vail, devised the Morse code by assigning the most frequently used letters the simplest symbols. By 1845, 12 words per minute were possible using paper tape on the Morse Telegraph. By the 1920s multiplexing was in use and speeds of 200 words per minute were being transmitted.

Progress was initially slow but by the 1850s the telegraph was an indispensable part of the railway system. In 1851 a telegraph cable was laid between England and France to connect to the rapidly developing European telegraph systems. By 1858 the *Agamemnon* and *Niagara* cable ships had laid a telegraph cable across the Atlantic but communication only lasted three months. In 1865 a second attempt was made but the cable broke. Success came in 1866 when the *Great Eastern* not only laid a new cable but retrieved the broken cable and two cables became operational between Britain and the USA.

The British Empire was expanding and communication was essential. By 1902, with the installation of the Pacific cable, the telegraph cable extended around the world and linked the outposts of the Empire. These telegraph cables survived until the 1950s when submarine telephony cables became available.

In 1876, in Boston, USA, a Scottish immigrant named Alexander Graham Bell succeeded in sending speech from one room to another using a variable resistance microphone and an electromagnetic receiver. On 7 March Bell beat Elisha Gray of Chicago by 3 hours in the race to patent the telephone. Both telephone systems were similar but Bell was granted the patent and this was then held by the United Telephone Company. The commercial version of Bell's telephone consisted of an electromagnetic receiver and transmitter but it was a carbon microphone, patented in 1878 by the Rev Henry Hunnings from Yorkshire, which eventually formed the basis of the modern telephone.

The telephone was quickly accepted in the USA but not in Britain where those who could afford it considered it an intrusion and an abundance of servants and messenger boys made it unnecessary. In 1879 the first manual switchboard was opened in London with eight subscribers. By 1883 there were still only 10 000 telephones in Britain. During the 1880s many different companies were competing for the telephone business, causing difficulties when connecting between different systems. By the early 1890s amalgamations took place and the National Telephone Company (NTC) took control of the telephones in Britain except for the few owned by the Post Office.

Connections between telephones were made by operators but in 1891 an undertaker named Almon Brown Strowger from Kansas patented the automatic telephone exchange and the first exchange was built in La Porte, Indiana. The first Strowger was built in England at Epsom in 1912 by the Post Office, who in that year had taken control of all the National Telephone Company exchanges. (The Post Office had already possessed the telegraph monopoly since 1868 and was given control of all the telephone trunk lines between towns in 1896 in order for the government to keep control of the new industry.) In 1918 Strowger was adopted by the Post Office for a programme of building more exchanges.

The London automatic system was opened in 1927 as a director system using three digits to select the exchange and a four digit subscriber number. Over 100 exchanges were planned and each exchange could connect 10 000 customers.

The early long-distance cables suffered loss and distortion and this was partially solved in 1900 with the invention of loading coils by Pupin and Cambell who were working independently in America. This meant placing inductances at specific intervals along the line and this removed some of the distortion caused by the loss of the high frequencies of the sound. At this time there was no amplification of the signal.

J.J. Thomson discovered the electron at the turn of the century and soon afterwards J.A. Fleming patented the diode. In 1906 Lee de Forest introduced the control grid between the anode and cathode of the diode and produced the triode valve and electronic amplification. This had major effects on cable telephony and radio. For the first 15 years of radio there was no means of amplification. Radio waves were now generated without the use of spark transmitters and receivers were manufactured with diode detection and triode amplification. Radio telephony became possible and was used during the Great War 1914–18. In 1927 a long wave telephony transmitter was built at Rugby by the Post Office to work with an American station at Rocky Point in New Jersey. As will be later seen these transmissions are inefficient and it was discovered that short waves (Chapter 2) could travel the same distance with less power and with the use of smaller aerials which could be made directional. By 1927 the Post Office was beaming a telegraph system around the world. This was known as the Empire Beam and by the 1930s it had expanded to include a telephony service. The service became the mainstay for world-wide communication for over 30 years.

The Marconi Company transmitted the first public broadcast dedicated to news and entertainment from Chelmsford on the 23rd. February 1920. After only a short period of time the Post Master General decided to close the station without any logical explanation. New stations in the USA, France and Holland were opening and listeners were able to receive continental programmes but for over a year there were no British transmissions. Public opinion was strongly in favour of the broadcasts and, after two petitions to the Post Master General, the Marconi Company was again allowed a limited public broadcasting service. The press opposed the broadcasting of news and entertainment as they considered it would affect their business. With the new service the number of listeners quickly increased and soon the interest generated allowed Marconi to ask the Post Master General for a licence for another station to broadcast a regular programme from London. With the aid of other radio manufacturers the Marconi Company formed the British Broadcasting Company (later Corporation) and the 2LO station.

The BBC started broadcasting on 14 November 1922 from the seventh floor of Marconi House in The Strand in London. On the second night two regional stations were opened in Birmingham and Manchester. The quality and quantity was variable but soon after the BBC moved to Savoy Place (now the home of the Institution of Electrical Engineers) before finally moving to Portland Place in 1932.

Cable telephony benefited from radio technology as signals were multiplexed to improve carrying capacity. In 1912 repeaters using valves were used to amplify signals on trunk routes and in 1927 H.S. Black devised negative feedback to improve telephony amplifiers and later this was adopted in all quality amplifiers (Chapter 4). In the late 1920s, as systems became larger and more complex, the problems of traffic density and equipment requirements for various qualities of service became a mathematical problem and such pioneers as A.K. Erlang, who gave his name to the unit of traffic, made important contributions to this knowledge.

In 1936, on 2 November, the BBC opened the first television broadcasting system using both the Baird mechanical system and the Marconi-EMI electronic system. After experiments the Baird system was discontinued and television continued with the electronic scanning system still used today although with a greater number of lines.

The war years of 1939 to 1945 saw the temporary closing of the BBC television service and a termination of the development of Europe's telecommunication network. Civil telecommunications were at a stand-still until after the war. However, the development of radar produced new knowledge and equipment for use in the UHF and SHF frequency bands and special valves, waveguides and insulating materials were developed. The army No. 10 radio sets were the first to use short wavelengths, pulsed modulation and time division multiplexing. Advances in marine, avionics and military mobile communications and navigation were made. The German V2 rocket technology led after the war to missiles and the launching vehicles for communication satellites.

Ten months after the end of the war the BBC reopened the television service in the London area but it was not until 1949 that the Midlands transmitter was opened at Sutton Coldfield. By 1953, at the time of the Coronation, 80% of the country could receive the service.

In 1947 John Bardeen, Walter Brattain and William Shockley invented the transistor at the Bell Telephone Laboratories. This was a major milestone for electronics and all three physicists received the Nobel Prize for physics in 1956. From individual transistors to complete circuits to microprocessors, the technology continually improved and integrated circuit manufacturers today are continuing to increase the packing density of components on a single piece of silica, making possible the equipment we are using in every aspect of our daily lives.

The first transatlantic telephone cable, TAT–1, was operational in 1956 and provided 36 telephone circuits. Between 1956 and the present day additional transatlantic cables have been laid from Britain, France and Spain, each one with a greater capacity than the last. Today a transatlantic cable carries thousands of multiplexed circuits and the latest contain fibre optics.

The date of 11 July 1962 saw the first live pictures transmitted by satellite from the USA to Britain. The satellite was Telstar and the pictures were received in the UK at the Post Office earth station at Goonhilly in Cornwall. The satellite was in low orbit and each orbit lasted $2\frac{1}{2}$ hours. The satellite was visible to Goonhilly for 20 minutes in each orbit and the receiving dish had to follow the satellite from one horizon to the other. It was 1965 before the first commercial geostationary orbit satellite, Early Bird, was launched by the INTELSAT organisation (see Chapter 2). Throughout the next years satellites were launched for communications, broadcasting, military purposes, navigation, surveying, weather forecasting and to study the environment. Man landed on the moon in 1969 and many probes have been sent to the planets. In 1964 Mariner 4 sent back 22 pictures from Mars and took 9 days to transmit them. By 1981 Voyager 1 and 2 were returning detailed pictures of Saturn at a rate of one every 48 seconds. This was made possible by improvements in coding and information transmission technology.

The 1960s saw two further major advancements. Integrated circuits allowed computers to begin a rapid development stage which in a short time revolutionised the method of control for almost all technical systems. The computer altered the business world and shortened the development time of complex systems with its ability to simulate designs and make rapid calculations. As computers began to be able to handle more and more data in a shorter time, analogue systems began to be phased out in favour of digital systems. We are still in this transition stage at the present time.

In 1966 a major development at the Standard Telecom Laboratory led to the fibre optic cable being produced. This was proposed by K.C. Kao and G.A. Hockham. In 1977 the first fibre link was installed between Hitchin and Stevenage and today the British Telecom trunk network is almost entirely fibre optic. The digital revolution in telecommunications will eventually see the end of analogue exchanges and the network will handle all signals in the same manner irrespective of whether they are speech, data, music, video, etc. The first System X digital exchange was installed and tested in 1979 and these or similar digital exchanges are now rapidly replacing analogue exchanges.

Private mobile radio has been rapidly expanding since the 1950s and paging systems were also first installed at this time in the UK. By the 1980s integrated circuits and computer technology allowed mobile radio systems to develop rapidly. All types of systems started to become available both locally and nationally. These included the public cellular systems (Cellnet and Vodafone) for the transmission of speech and data, comprehensive paging systems and trunked networks which allowed PMR channels to be pooled and the available spectrum to be used more efficiently. The 1990s will see the systems developing to become international and digital. All technologies will be used and it must be realised that most communication systems are interdependent and, irrespective of the final connection to the user, the technologies of other parts of the system are just as important.

The short history provided in this chapter is obviously only a small part of the immense work by thousands of people who have contributed to today's systems. Technology is so complex today that teams of people throughout the world rather than individuals make the major contributions to research and development. It is still important, however, that the student has some knowledge of how the industry developed and the incredible people who had the ability, enthusiasm, inquisitiveness and vision to make the initial developments on which everyone else builds.

2 Electromagnetic waves

Electromagnetic waves occur both naturally and as the result of electrical generation. When purposely generated as radio waves they become the carriers of
information. Sound and vision will not travel far in their original state. Sound will
only be heard depending upon the loudness of the voice and vision will be blocked
by any opaque object. If these signals are to travel long distances they must be
changed into electrical signals and combined with a signal that has the ability to
travel long distances and pass through obstructions. Electromagnetic waves have
characteristics which are determined by their frequency and, therefore, the choice
of carrier wave for our communication signal will determine its transmission quality, the distance it can travel, the ease with which it can be transmitted and received
and the places in which it can be received.

Whenever an electrical signal passes through a conductor there will be associated
with it a magnetic field at right angles to the conductor. If the electrical signal is
direct current the magnetic field around the conductor will be static. If, however,
the current is alternating the magnetic field will build up in one direction, collapse
and build up in the opposite direction in sympathy with the alternating current.

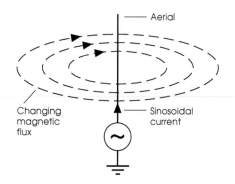

Figure 2.1 *When a sinusoidal current flows in
the conductor an alternating magnetic field
is produced around it. Above approximately
10 KHz electromagnetic waves are propagated. As the current changes the magnetic
field collapses and builds up in the opposite
direction.*

Radiation of an electromagnetic wave

When the magnetic field is changing and the magnetic flux cuts the conductor an electric field is induced in the conductor. (This is, of course, the principle by which we generate our electricity supply. Electrical conductors are rotated in a magnetic field to induce the voltage.) The electric and magnetic fields of an alternating current are, therefore, dependent upon each other.

There is a finite time necessary for the magnetic field to build up, collapse and build up in the opposite direction and for frequencies above approximately 10 kHz there is insufficient time for all the energy to return to the conductor before the current has reversed and begins to build up the magnetic field in the opposite direction. The energy which has not returned is radiated from the conductor as an electromagnetic wave. These waves have a constant speed of 300 000 000 metres per second (186 000 miles per second). As light waves are also part of the electromagnetic spectrum this is the speed of light. (This should be compared to the speed of sound which is approximately 760 miles per hour, depending upon the characteristics of the medium it is passing through). The electromagnetic waves, unlike sound waves, will pass through a vacuum. If they did not we would be unable to see the stars or communicate with satellites.

Polarisation

By convention we state the polarisation of an electromagnetic wave by reference   to the electric field (see Figure 2.2). Whether it is a vertically or horizontally polarised signal depends upon the direction of the electric field. As it is the action of the magnetic field cutting the receiving aerial (conductor) which induces the received signal, it is essential that the aerial is in the correct plane to the radiated signal otherwise a signal either will not be received or will be considerably reduced in strength.

Frequency and wavelength

Instead of stating the frequency of a radio wave we often refer to its wavelength. This avoids using large numbers when frequencies are high. It also relates the signal

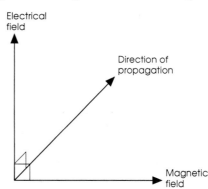

Figure 2.2   *The electric and magnetic fields exist at right angles to each other and both are at right angles to the direction of propagation. By convention the radio wave is vertically polarised if the electric field is vertical. The positions of the electric and magnetic fields are reversed for a horizontally polarised wave.*

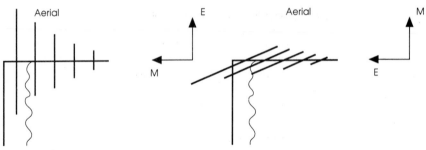

For reception of vertically polarised
radio waves

For reception of horizontally polarised
radio waves

Figure 2.3   *In order to induce a signal voltage the receiving aerial must be cut by the magnetic field of the radio wave and must, therefore, be at right angles to the magnetic field for
maximum signal.*

directly to the size of aerials, dishes, transmission lines and waveguides for efficient
transmission and reception.

The wavelength is directly proportional to the velocity of the wave and inversely
proportional to the frequency.

In mathematical terms:

$$\text{Wavelength } (\lambda) = \frac{\text{velocity}}{\text{frequency}} = \frac{300\ 000\ 000}{\text{frequency}} \text{ metres}$$

$$\text{Frequency } (f) = \frac{1}{\text{period of time for one cycle in seconds.}} \text{ hertz (Hz)}$$

Examples
1. The frequency of a signal which takes 20 ms for a complete cycle is:

$$\text{Frequency } (f) = \frac{1}{20} \times 1000 = 50 \text{ Hz}$$

2. A radio frequency of 200 kHz has a wavelength of:

$$\text{Wavelength } (\lambda) = \frac{300\ 000\ 000}{200\ 000} = 1500 \text{ metres}$$

Electromagnetic wave characteristics

The particular characteristics and, therefore, the uses of the electromagnetic waves
depend upon their frequency. The range of frequencies of the electromagnetic
waves is shown in Figure 2.4.

Cosmic rays have great penetrating power and reach the world from outer space
but their origin is unknown. Some cosmic rays are believed to originate in the sun.

Gamma rays (10^{-10} to 10^{-14} metres) are released by radium and other
radio-active materials. They are used in hospitals to kill cancer cells and other
body diseases but their sources must be kept in thick lead containers for safety
purposes.

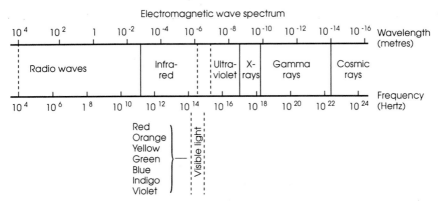

Figure 2.4 *X-rays range from 10⁻⁹ m to 10⁻¹¹ m. The shorter wavelengths are known as hard X-rays and the longer wavelengths are known as soft X-rays. The transmission characteristics of radio waves depend upon their frequency.*

$$\text{Wavelength in metres} = \frac{\text{velocity}}{\text{frequency}} = \frac{3 \times 10^8}{f \ (\text{Hz})} \ \text{m/s}$$

X-rays (10^{-9} to 10^{-11} metres) are produced when heavy metal atoms (usually tungsten) are struck by sufficiently energetic electrons in a vacuum tube. They have the ability to penetrate the body tissues and reasonably thick metal. They are used as a diagnostic tool to produce X-ray photographs in the medical world, industry and science.

Ultraviolet rays (between approximately 380 and 5 nanometres) are present in sunlight and will kill life if doses are large although small doses have beneficial effects and are used in dermatology. The world is protected by an ozone layer in the atmosphere which absorbs the ultraviolet rays below 200 nm and prevents them reaching the earth. Ultraviolet rays are responsible for the creation of the ionosphere which is layers of charged electrical atoms of gases. These have a considerable effect on radio transmission. Ultraviolet radiation is produced during arc discharges and by gas discharges such as a mercury vapour lamp.

Visible light (740 nm to 400 nm) occupies about one-seventieth of the electromagnetic spectrum. Our eyes receive the electromagnetic waves in this part of the spectrum and produce electrical impulses which are interpreted as vision by our brain. Each of the visible colours occurs as a different frequency (red, orange, yellow, green, blue, indigo and violet). White light consists of a mixture of all the colours. Other creatures have the ability to see in different parts of the spectrum such as ultraviolet.

Infrared (750 nm to 1 mm) is experienced as heat. It can penetrate the body and generate internal heat. Objects above absolute zero emit infrared rays and these can be detected by infrared detectors for night viewing and alarm systems. It is the predominant radiation from a body up to a temperature of about 3000 °C.

Below the frequencies of the infrared rays are the microwaves and radio waves used for communication systems and heating effects. Their communication uses appear later.

Ionosphere

As we rise in height above the earth the atmosphere becomes rarer and ultimately a vacuum. Ultraviolet light from the sun enters the atmosphere and provides sufficient energy to ionise the molecules of gases by removing some of the electrons from their atoms. The free electrons may recombine with a positive ion and form a neutral atom while others remain free to produce an ionised layer of gases. Nearer the earth the air is denser and the energy of the ultraviolet light is absorbed. The possibility of a recombination of the atom is also greater and these factors prevent ionisation layers from occurring. The density of the free electrons, therefore, varies from very low at high altitudes above 500 km, where air molecules are rarer, to high density at lower altitudes between 50 and 500 km and low again above the surface of the earth.

The ionised area exists in layers which, because they are caused by the ultraviolet rays from the sun, vary from summer to winter and night to day. The layers, starting from the closest to the earth, are given the designations D, E, F_1 and F_2. At night the D layer disappears and the E layer becomes weaker. The F_1 and F_2 layers also merge and their height will be very variable.

Unless an aerial has been specially designed to be directional it will radiate both ground waves and sky waves. The attenuation of these different signals and the transmission of the sky wave will depend upon their frequency. The ionosphere layers have a considerable effect on the transmission of sky waves below a frequency of 30 MHz. As the electron density of the ionosphere increases with height it causes the refractive index to decrease. When radio waves enter these layers they are refracted away from the normal and this effect is more apparent as the radio frequency is reduced. Radio waves below 30 MHz which enter these layers are, therefore, returned to earth as sky waves. Above 30 MHz the radio wave is insufficiently refracted to cause it to return to earth and it is able to pass through the ionosphere. There is, therefore, a need to use very high frequencies for

Figure 2.5 *The sun's rays cause the ionisation of the upper atmosphere. This was first suggested by A. F. Kennelly and independently by Oliver Heaviside in 1902. Edward Appleton proved their existence by bouncing radio waves off the different layers. (The F region is known as the Appleton layer and the lower E levels are known as the Kennelly–Heaviside layer.) The layers change between day and night and time of year. Sun-spot activity also affects the ionosphere and radio transmission.*

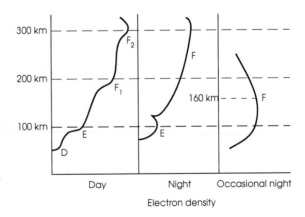

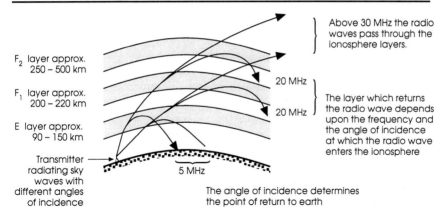

Figure 2.6 *The diagram shows the ionosphere when the seperate E, F₁ and F₂ layers exist. Two angles of incidence are shown for three different frequencies.*

communication with satellites which are above the ionosphere.

The actual layer of the ionosphere which returns any particular frequency to earth will also depend upon the angle of incidence at which the radio wave enters. The smaller the angle from the normal that the wave enters the layer the greater must be the refraction before it is returned to earth. Waves of the same frequency can, therefore, reach higher layers of the ionosphere before refraction returns them to earth if the angle of incidence at which they enter the ionosphere is reduced.

Skip distance

The distance between the end of the useful ground wave and the first position at which the sky wave can be received is known as the skip distance. There is no reception of the signal between the end of the ground wave transmission and the returned sky wave. Skip distances can be used for long-distance communication and

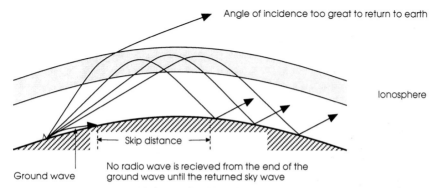

Figure 2.7 *The radio waves reflected from earth will return to the ionosphere and will again be reflected back to earth. Multiple hop links allow high frequency communication to any part of the world.*

practical distances of approximately 2000 km can be obtained from the E layer, 3000 km from the F_1 layer and 4000 km from the F_2 layer. Multiple hop transmission is possible in order to provide transmission around the world. This requires reflection of the radio signal from the earth back to the ionosphere where again refraction or reflection takes place. Very low frequencies do not penetrate the ionosphere and are, therefore, reflected from the base of the ionosphere layer.

Multiple hop transmission can, however, be erratic and when used for reliable long-distance communication a number of receiving stations are used and switched between in order to obtain the best reception. For continuous communication using sky waves the frequency and angle of incidence are changed with changing conditions in the ionosphere.

At night a listener to the radio will notice an increase in the reception of foreign stations. This is due to the lower layers of the ionosphere disappearing and the radio waves being refracted from higher layers of the ionosphere and, therefore, travelling further.

Critical and maximum usable frequency

If sky waves are to be used for transmission it is important to establish the critical and maximum usable frequency. The critical frequency for any layer of the ionosphere is the highest frequency that can be transmitted vertically upwards and received back on earth after refraction from the layer.

The maximum usable frequency (MUF) is the highest frequency that can be used to transmit and receive a sky wave at particular points on earth. Because a radio wave is attenuated more at lower frequencies by the ionosphere (attenuation $\propto 1/f$) there is an advantage in using the highest possible frequency which maintains continuous communication. In practice because the ionosphere is continually changing an optimum working frequency is used which is approximately 80% of the maximum.

The maximum usable frequency is dependent upon the angle of incidence at which the radio wave enters the ionosphere. It is related to both the critical frequency and the angle of incidence by:

$$\text{MUF} = \frac{f_{\text{critical}}}{\cos (\text{angle of incidence})}$$

and $f_{\text{critical}} = 9\sqrt{N_{\text{max}}}$

where N_{max} is the maximum electron density of the layer in the ionosphere.

Satellites can be regarded as an artificial ionosphere whose position and transmission specification are accurately known and controlled and, therefore, provide more reliable long-distance transmission than the ionosphere.

Frequencies used for radio communication

The frequencies used for radio communication can be divided into five bands in order to study their propagation:

1. Below 100 kHz
2. 100 kHz to 1500 kHz
3. 1500 kHz to 6 MHz
4. 6 MHz to 30 MHz
5. Over 30 MHz

Below 100 kHz transmission is by ground waves or multiple reflections between the ground and the base of the ionosphere. Attenuation is small but, because of the large aerial systems required for these low frequencies, the transmitters are high powered and very inefficient. They are satisfactory for very long distances where communication is continuously required. Because the radio waves do not penetrate the ionosphere their reliability is independent of time of day or season of the year. The level of static on these services can be high and because of the distances these waves can travel only a few transmitters can be accommodated in the available frequency spectrum. They are used for broadcasting, ship-to-shore communication and some navigation systems.

100 kHz to 1500 kHz

At the lower frequencies the signal will be propagated mainly as a ground wave as the sky wave is highly attenuated, especially during the day, and distant stations may only be received at night. These frequencies are mainly used for broadcasting, direction finding and certain navigational aids.

1500 kHz to 6 MHz

These have similar characteristics to the previous group with the exception that daytime attenuation is less. Both ground wave and sky wave are used for transmission. This band is used for moderate distance communication and direction finding.

Figure 2.8 *When a ground wave is transmitted the wavefront is tilted because the radio wave does not travel as quickly through the earth as in space due to energy losses in the earth. The tilting allows the wave to*

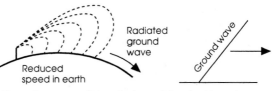

follow the curvature of the earth. Ground waves are the major type of signals at long waves. Medium waves travel as both sky and ground waves. Ground waves must be vertically polarised otherwise the electric field would be short circuited. Attenuation of low frequency ground waves depends on distance, frequency and earth conductivity. There are very low losses over sea, but they are high over dry land.

Figure 2.9 *Low frquency radio waves not only travel as ground waves but also can reflect from the base of the ionosphere and travel considerable distances as a multi-hop transmission. Losses are low on each reflection and the received signal strength is inversely proportional to distance travelled.*

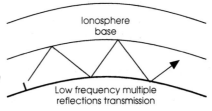

6 MHz to 30 MHz

In this band the transmission depends almost entirely on the sky wave and, therefore, the conditions in the ionosphere will greatly affect the attenuation and the distances travelled. However, when the conditions are good long distances can be covered with little attenuation. This band is used for long-distance broadcasting and communication.

Above 30 MHz

These radio waves pass through the ionosphere and are used where a line of sight between the transmitter and receiver is possible although this path may also include reflections. These frequencies are used for broadcasting both radio and television, point to point radio links including transmission to and from satellites, radar and mobile communications (see Chapter 8 for more details). The short wavelengths of the transmitted signals allow small efficient aerials to be used which are a considerable advantage for mobile communications. These signals travel relatively short distances when compared to lower frequencies, unless highly directional dishes are used in line of sight to satellites, and generally no more than 40 km (25 miles). This is determined by the curvature of the earth. Frequencies can, therefore, be reused within relatively short distances without transmitters interfering with each other. Mobile radio systems extensively use the UHF and VHF frequency bands (see Chapter 7) and, in order to maximise the number of channels in the available spectrum, the channel frequencies are constantly reused. Transmission distances, therefore, have to be controlled and this is achieved by controlling the radiation pattern produced by the antenna (Chapter 8) and by adjustment of the radiated power from the transmitter. These are both important when designing mobile radio systems as precise areas must be covered.

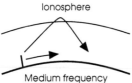

(a) Ground Wave (see also Figure 2.9).

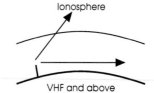

(c) Sky waves – long-distance communication by multiple hops.

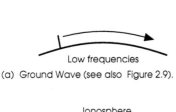

(b) Ground wave and sky wave. The D layer of the ionosphere causes considerable attenuation but disappears at night. Maximum attenuation at 1.4 MHz.

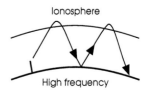

(d) Line-of-sight transmission. Radio wave is not returned by the ionosphere.

Figure 2.10   *Summary of radio wave transmissions.*

Classification of radio frequencies

The allocation of radio frequencies is strictly controlled and monitored throughout the world (see Chapter 7). If radio transmission is to be successful, interference by unauthorised transmitter users must be prevented.

The radio frequency spectrum is divided into bands and these are referred to as follows:

10 kHz to 30 kHz	VLF	Very Low Frequency
30 kHz to 300 kHz	LF	Low Frequency
300 kHz to 3000 kHz	MF	Medium Frequency
3000 kHz to 30 MHz	HF	High Frequency
30 MHz to 300 MHz	VHF	Very High Frequency
300 MHz to 3000 MHz	UHF	Ultra High Frequency
3000 MHz to 30 GHz	SHF	Super High Frequency
30 GHz to 300 GHz	EHF	Extra High Frequency

(1000 MHz = 1 GHz).

The UHF and SHF bands have been further divided:

UHF	300 MHz to 1000MHz
L band	1000 MHz to 2000 MHz
S band	2000 MHz to 4000 MHz
C band	4000 MHz to 8000 MHz
X band	8000 MHz to 12 500 MHz
Ku band	12.5 GHz to 18 GHz
K band	18 GHz to 26.5 GHz
Ka band	26.5 GHz to 40 GHz
millimetre	> 40 GHz

(Source: IEEE Standard 521–1976)

Fading and multipath transmission

It has been seen that medium frequencies can travel as both a ground wave and a sky wave and arrive at the receiver after having travelled different distances. The received signal in such instances is a phasor addition of both signals and depending upon the

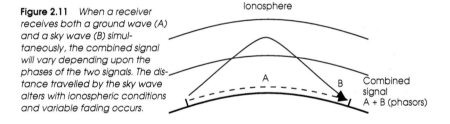

Figure 2.11 *When a receiver receives both a ground wave (A) and a sky wave (B) simultaneously, the combined signal will vary depending upon the phases of the two signals. The distance travelled by the sky wave alters with ionospheric conditions and variable fading occurs.*

actual phases of the signals they may add or subtract from each other. This produces a variable signal level at the receiver. In order to minimise this type of fading the transmitter aerials are designed to maximise the ground wave and minimise the sky wave.

In the high frequency range of radio signals, the transmission is by sky waves but the received signal may be the addition of two or more paths through the ionosphere and this problem can be multiplied when the transmission path relies on more than one hop. As the ionosphere is continually varying the phase differences will change and produce an erratic signal.

The variation of the signals is dependent upon the frequency and, therefore, different parts of the received signal can be affected in different ways. This is known as selective fading and becomes a problem particularly when double sideband transmission is used (see Chapter 3). Selective fading cannot be compensated with automatic gain control as this operates by reference to the carrier which may or may not be affected.

The problem can be minimised by use of frequency modulation for the transmitted signal, using a carrier close to the MUF and the careful design of the transmitting aerial in order to radiate only one mode of propagation.

Any radio wave travelling through the ionosphere will be attenuated and this factor is variable due to the variations in the conditions within the ionosphere. This is known as 'general fading' and can sometimes lead to the complete loss of the signal. As it is the complete signal in this instance which is varying, automatic gain control in the receiver, which adjusts the gain of the receiver dependent upon the strength of the received signal, can improve the reception.

Multipath transmission also occurs in the VHF and UHF bands as these signals are reflected by hills, buildings and moving objects. This also means that receiving aerials can be screened by objects and topography and be unable to receive a signal. When reflections occur, again the received signal will vary depending upon the addition of the phases of the different signals. These will vary depending upon the distances travelled before arrival at the aerial. If the receiver or reflecting object is moving this will produce variable fading. It is, therefore, important to provide a strong signal with a direct path to the receiving aerial if these problems are to be minimised (see Figure 2.13).

Figure 2.12 *Due to the different angles of incidence two signals A and B can be received which have taken different paths through the ionosphere. Signals A and B will add or subtract depending on their phase difference.*

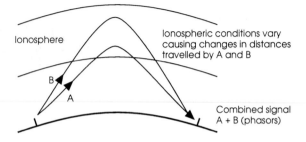

Ionosphere

Ionospheric conditions vary causing changes in distances travelled by A and B

B

A

Combined signal A + B (phasors)

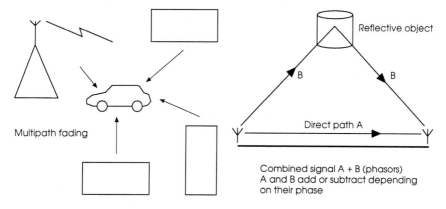

Reflective object

B   B

Direct path A

Multipath fading

Combined signal A + B (phasors)
A and B add or subtract depending
on their phase

Figure 2.13 *In the UHF and VHF range of transmission frequencies a mobile can receive signals transmitted directly and reflected signals from buildings and hills. The reflected signals will have travelled different distances and the combined signals at the mobile can add or subtract and result in fading. When a mobile is moving, the received phases from the reflected signals will change and the fading is variable. This is known as Rayleigh fading.*

If the transmitted signal is digital, the time taken for the different signals to arrive at the receiver can cause the digital 1 and 0 bits to be distorted and the receiver is unable to identify the signal correctly. This is known as time dispersion. This type of distortion affects the bit rate that can be used as a reflected digital 1 may be received when the direct signal is 0. (See Chapter 7 for digital systems.)

Tropospheric scatter propagation

The troposphere is the lower levels of the atmosphere nearest the earth. In this region the temperature falls with increase in height and is the region in which most meteorological phenomena occur. It varies in thickness from about 7 km (4.5 miles) to about 50 km (31 miles) at the equator. If a radio wave above 600 MHz is transmitted upwards from a high power transmitter, some of the energy is scattered by the troposphere in a forward direction. This has to be received by a high gain aerial but a reliable long distance signal can be received over a range of 300 to 500 km.

Figure 2.14 *Forward scattering of UHF signals by the troposphere can be detected at distances of approximately 500 km or even further. The received signal is subject to both rapid and slow fading which increases with distance. In 1976 a telephone service to the North Sea oil platforms was inaugurated using a microwave radio link and tropospheric scatter.*

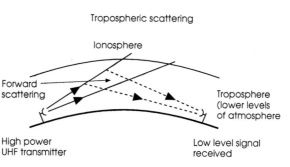

Tropospheric scattering

Ionosphere

Forward
scattering

Troposphere
(lower levels
of atmosphere

High power
UHF transmitter

Low level signal
received

Ionospheric scattering can also be used as the lower ionosphere is always turbulent. VHF frequencies are used and a high power transmitter is required. Signals can be detected at distances in the range 1000–2000 km.

Although inefficient due to the transmitter power that is required for the reception of a small signal at the receiver, it has uses where the terrain makes other communication systems difficult or impossible.

Satellites

Satellites have the advantage over other forms of communication systems in that anyone who has line of sight to a satellite and the correct equipment can transmit and receive signals without further connections over all the area visible to the satellite. This, of course, assumes the transmitter has access to a transponder on the satellite. The transponder receives the signals transmitted to it, amplifies them and retransmits them on a different frequency. The satellite transmitting aerial may be directional, as in the case of direct broadcasting satellites, so that the transmission is aimed at specific areas covered by the satellite. This allows individual language broadcasts to be targeted to specific countries.

The power for the equipment is obtained from banks of solar cells which are photovoltaic. If the satellite is spinning only part of the array is pointing to the sun at any particular time whereas a non-spinning satellite has wings which can be pointed to the sun at all times and this allows maximum efficiency. Some spinning satellites have been designed with additional wings attached to their non-spinning section in order to increase power.

Orbits

Satellites can be placed in orbit at any height above the earth but because of drag and heating this orbit is outside the atmosphere. A satellite moving in a circle around the earth is attracted towards the centre of the earth by gravity. There is, however, also a tangential force on the satellite due to its speed. At any particular height there is a speed where the two forces balance and the satellite stays in the orbit. This speed is called the orbital velocity.

At a height of 35 786 km (22 237 miles) a satellite requires a speed of approximately 7000 miles per hour (11 300 km/h) in order to stay in orbit and this speed takes it round the earth in 24 hours. If the satellite is going eastwards in the same direction as the earth and it is exactly over the equator it will appear to stay over the same point on earth. A tracking station will, therefore, see the satellite in the same position in the sky at all times. This is known as geostationary orbit and a satellite in this position is capable of covering 40% of the earth's surface. Three such satellites spaced 120 degrees apart can cover nearly all the world except some polar regions. Earth stations within sight of the same satellite can communicate with each other directly through the satellite. Earth stations which see different satellites can communicate via ground stations which can see both satellites. These stations receive the signal from the first satellite and then retransmit to the second satellite. Direct transmission between satellites is not performed at present.

Geostationary orbits are normal for communication satellites and direct broad-

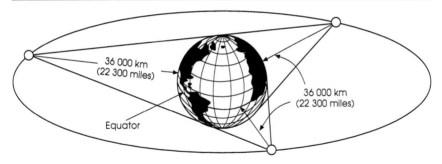

Figure 2.15 *A synchronous satellite takes 24 hours to circle the earth in a circular equatorial orbit. It therefore appears to remain stationary over a fixed point on the equator. The unique position of 36 000 km above the equator allows three synchronous satellites to provide almost complete coverage of the earth. This geostationary orbit is more precisely 35 786 km (22 237 miles).*

casting satellites. However, other orbits are chosen for different purposes. Surveying of the earth's surface and weather (metsats – meteorological satellites) is done with low polar orbits and as the earth spins the satellite is able to record a different strip of the earth on as many as 14 orbits each day. In addition there are five geostationary metsats, each viewing approximately one quarter of the earth's surface, which continuously record weather changes throughout the world. Spy satellites use a long narrow ellipse for an orbit. This allows the satellite to swoop low at maximum speed to look at the earth before moving away to a much greater distance from the earth.

Navigation by satellite systems

Satellites have become important for navigation and these systems use a number of satellites in different orbits and at any time several are visible to an observer. Knowing the position of these different satellites at any particular time gives the observer their exact position. The American Navstar system, also known as The Global Positioning System (GPS), became fully operational during 1993. Although it is designed and developed as a US military satellite system providing world-wide navigation capability, it has many other applications, notably in offshore and engineering surveying, geodesy and geophysics. Its accuracy in practice has exceeded expectations and its capability has been limited for all those people without special access to the satellites. The accuracy is approximately 15 to 25 metres for military users and 100 to 150 metres for civilian users. However, in scientific and commercial applications, with the aid of special observational techniques and computational algorithms, accuracies of a few centimetres or even millimetres are being achieved. There is a problem, however, in that the positions do not relate directly to many current maps due to the datum lines and projections used by map makers (the standard for GPS maps is WGS 84).

The system uses 24 satellites in six different orbits at a height of 20 000 kms (12 428 miles). These circle the earth every 12 hours and at any time an observer has four satellites visible from any point on earth. The satellites carry very accurate

Figure 2.16 *Global positioning system (GPS). GPS uses 24 satellites (25 launched) arranged in 6 orbits. The atomic clocks in the satellites are synchronised and all satellites simultaneously send 2 radio waves in the L band (1574.2 MHz and 1227.6 MHz) which carry coded information stating the time the signal left the satellite and its position. By sending 2 radio waves at different frequencies it is possible to compute errors due to ionospheric effects affecting the speed of radio waves (lower frequencies are slowed more).*

Receivers compute their position by comparison of the time the signals are received and the satellite positions. There is also a similar Russian system (GLONASS).

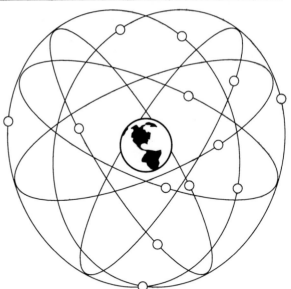

atomic clocks and simultaneously transmit two radio signals to earth. It is the differences in arrival times of the signals indicating a particular time when the signals were sent from the different satellites and the known positions of the satellites which allow the very accurate computation of position. The satellites are monitored from ground stations and both the atomic clocks and the satellites' positions are closely monitored and corrected. In addition to position, speed and height can also be measured. Small portable receivers are now available which indicate their position and these can be used by individuals carrying a lightweight pack.

Radio communication with satellites

Satellite communication requires frequencies which pass through the ionosphere to be used for transmission and reception and the common frequencies which are used for communication satellites are in the C, Ku and K bands (see the section on classification of radio frequencies). Signals are small and the use of dishes to maximise the received signal is necessary. There are advantages in using the higher frequencies as the dish can be smaller for transmitting and as the proposed minimum spacing of satellites is to be 2 degrees the transmitting dish will have to have a diameter of at least 100 wavelengths to avoid interfering with adjacent satellites.

The frequencies used are subject to attenuation by rain and snow and signal-to-noise ratio deteriorates under these conditions. A clear path is essential for good reception and trees and similar objects can obliterate the signal.

When receiving, the noise and the bandwidth of the system is important. The smaller 1–3 metre diameter dishes cannot provide a narrow beam and may pick up radiation from other satellites, which appears as noise. To provide sufficient signal-to-noise ratio a large bandwidth is required from the frequency modulated signal (see Chapter 5) and this is achieved by increasing frequency deviation. (Refer to Chapter 7 for satellites for mobile radio.)

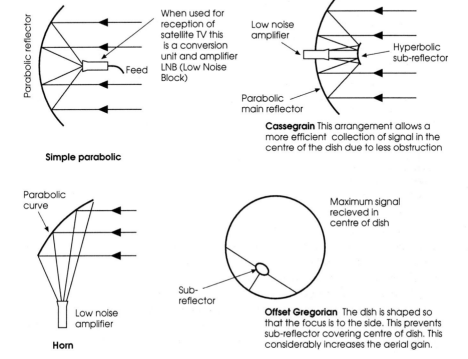

Microwaves are focused and concentrated both for transmission and reception by the use of parabolic reflectors. Received signals are focused at the parabolic focal point and if the receiving antenna is placed at this point the energy is concentrated. The head amplifier is usually placed at this position and signals can be converted to lower frequencies for transmission on the feeder cable to reduce losses.

If the transmitting antenna is placed at the focal point, parallel beams of radio energy are transmitted.

Accurately alined transmitter and receiver dishes are used for terrestrial communication links for distances of approximately 25 miles (40 km). Longer distances use multi-hop links. The curvature of the earth prevents transmission over longer distances.

Figure 2.17 *Dishes used for satellite and terrestrial microwave transmission and reception.*

3 Modulation

In the previous chapter it was shown that radio waves have the ability to travel long distances depending upon their frequencies and the output power of the transmitters. In order for them to become the carriers of the information there is a need to combine the required signal with the radio wave. In mobile radio the required signal is either audio speech or data. The same principles also apply to the transmission of television and sound broadcasting.

The original audio and visual signals are changed into electrical signals by microphones or cameras. Data will already be in the form of electrical binary digits which are produced by the computer. These electrical signals must then be combined with the radio waves before transmission. If the radio waves and information signals are only mixed together then the two signals will still exist as independent signals. The two signals must be combined in a process known as modulation in order for the radio signal and information to become one entity. Once the radio wave has been modulated with the information signal, the spectrum of the radio wave is altered and the information signal (modulating signal) ceases to exist as an independent signal. The two common methods of modulation are amplitude modulation and frequency modulation. Both systems are used extensively in broadcasting, line transmission and mobile radio. Phase modulation, which is similar to frequency modulation, is used in some data systems. Digital signals are usually encoded before the radio modulation process or transmission on a line.

Amplitude modulation

The carrier wave must be a very accurately produced frequency and preferably a crystal oscillator is used which has good stability. Additional precautions can be used to avoid drift due to temperature changes by placing the crystal in a temperature controlled oven.

The process consists of the amplitude of the carrier being varied in sympathy with the modulating signal (Figure 3.1). The carrier can be varied to any depth of

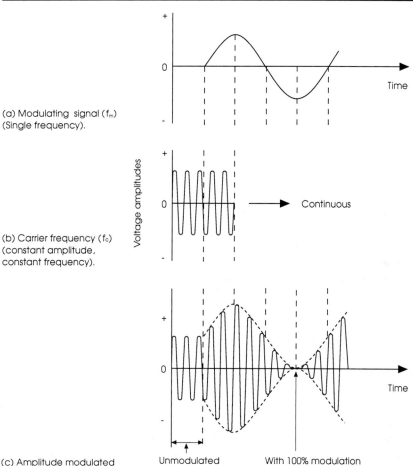

(a) Modulating signal (f_m)
(Single frequency).

Voltage amplitudes

Continuous

(b) Carrier frequency (f_c)
(constant amplitude,
constant frequency).

Time

(c) Amplitude modulated
waveform.

Unmodulated
carrier

With 100% modulation
the carrier is reduced
to zero

Figure 3.1

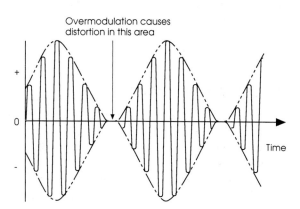

Overmodulation causes
distortion in this area

Time

Figure 3.2

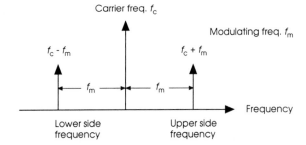

Figure 3.3 *Spectrum of an amplitude modulated double sideband waveform. (Single modulating frequency.)*

modulation that is required but it can be seen in Figure 3.2 that distortion occurs at minimum carrier level if the system is overmodulated.

When amplitude modulation takes place the frequencies of the modulating signal disappear and a new frequency spectrum is produced. This consists of two sidebands known as the upper sideband (USB) and lower sideband (LSB).

If the carrier frequency is f_c and the modulating signal f_m then USB $= f_c + f_m$ and LSB $= f_c - f_m$. The carrier frequency f_c also exists. The bandwidth required to transmit the two sidebands is $2 \times$ modulating frequency (Figure 3.3).

Modulating frequencies

In the examples a single modulating frequency has been shown for simplicity. In practice speech, music, video and data consist of bands of frequencies which must be transmitted without distortion in order to preserve quality and intelligibility. Typical bandwidths are:

Telephone conversation (commercial speech) 300 to 3400 Hz (Figure 3.4).
Human hearing 20 to 18 000 Hz but varies and is usually dependent on age.

Figure 3.4 *A telephone speech bandwidth of 300–3400 Hz (commercial speech) in an amplitude modulated double sideband system produces an upper and lower sideband. Within the sideband are all the frequency components produced with every frequency present in the modulating signal. Exactly the same information is present within the upper and lower sideband. One sideband can, therefore, be removed without losing any information.*

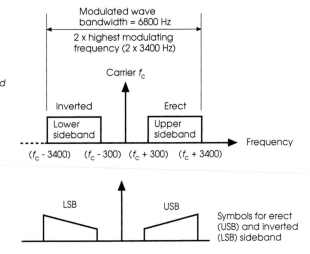

High quality music is the same as human hearing but is acceptable with restrictions; 20 to 10 000 Hz is good. Medium wave broadcasting is restricted to 4500 Hz as stations are spaced 9 KHz apart and are transmitted as double sideband transmissions.

Television pictures in PAL system require 5.5 MHz bandwidth for full definition but can be restricted to 1.5 MHz on domestic video recorders and be acceptable. Digital signals are dependent upon the bit rate and are often encoded before transmission in order to reduce the initial bit rate and, therefore, the bandwidth required.

Single sideband transmission

It is apparent from the frequencies within the upper and lower sidebands that the same information appears in both sidebands. If there is a need, therefore, to conserve the bandwidth of individual signals in order to increase the use of the available spectrum then one sideband can be suppressed and only one sideband transmitted.

Power in an amplitude modulated wave

The power in the sidebands is dependent upon the depth of modulation and hence the modulation index. If a carrier is modulated to a depth of 50% of its unmodulated level its modulation index (m) is 0.5.

The total power in an amplitude modulated signal is given by:

$$P_t = P_c \left(1 + \frac{m}{2}\right)$$

where P_t = total power and P_c = power in unmodulated carrier. If the maximum modulation is used (100%, $m = 1$), it can be seen that the combined power in both sidebands is only half that in the carrier. As it is only the sidebands which contain the required information this is an inefficient system.

Suppression of carrier and sideband

In order to increase power and spectrum efficiency the carrier or a sideband or both can be suppressed. These modulation systems add to the complexity of both the transmitter and receiver.

Double sideband suppressed carrier (DSBSC)

Suppression of the carrier in a double sideband system is performed by a balanced modulator of which there are several types. The output of such modulators consists of the upper and lower sidebands but the carrier is suppressed. However, the output waveform without the carrier is the resultant of the sidebands and the envelope is no longer sinusoidal for a sinusoidal modulating signal. In order to demodulate such a waveform it is necessary to reinsert the carrier at the receiver at both the correct frequency and phase. This requires an oscillator of very high stability at the receiver.

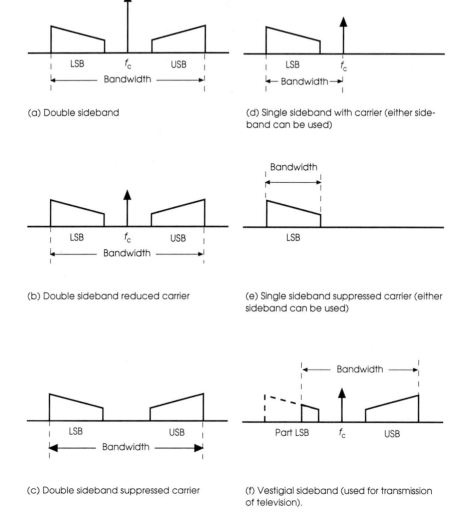

Figure 3.5   Different systems of amplitude modulation transmission. *Different systems are used to save bandwidth or power. When the carrier is suppressed it must be reinserted in the receiver and the frequency and phase must be very accurate.*

Single sideband suppressed carrier (SSBSC)

In order to increase the number of channels in a given radio spectrum, one sideband can be suppressed in addition to the carrier. This makes space for an additional channel. Again the carrier is suppressed by the use of a balanced modulator after which the unwanted sideband is removed with a bandpass filter. Before demodulation at the receiver the carrier must again be inserted although, in this instance, the phase is not important as only one sideband is present. However, the oscillator must be of very high stability if distortion of speech and especially data is not to occur. In some older systems a low level pilot tone was added to the transmitted signal which was used to lock the receiver's oscillator by means of an automatic frequency control circuit (AFC).

When using SSB transmission there is an improvement in signal-to-noise ratio when compared to DSB as the added noise is proportional to the bandwidth which is now halved. An additional improvement also occurs as the ratio of sideband power to the total power is also greater.

When selective fading occurs the sidebands of a DSB transmission may beat together and create unwanted frequencies if the carrier is reduced. This causes distortion. Such problems do not occur with SSB transmission.

Suppression of the carrier improves multi-channel systems which pass through amplifiers or other non-linear devices. Amplifiers are more susceptible to intermodulation distortion at higher levels of power and suppression of the carrier reduces this distortion.

Independent sideband amplitude modulation system (ISBSC)

In this system the outputs from two separate balanced modulators, operating on the same frequency and each producing a suppressed carrier, are passed through two separate bandpass filters. These produce sidebands equivalent to the upper and lower sidebands of an amplitude modulated waveform but in this instance they belong to different signals. The two separate signals are combined to produce a double

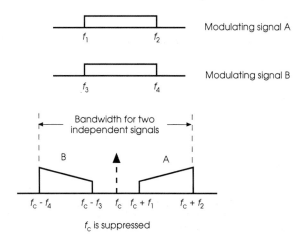

Figure 3.6   *Independent sideband amplitude modulation system. Two separate modulating signals have the same carrier frequency but occupy either the upper or lower sideband.*

sideband suppressed carrier waveform containing two separate signals. This system allows channels to be spaced closer together. A low level pilot tone can also be added to operate the automatic frequency control in the receiver where the carrier must again be reinserted in order to demodulate the signals.

Vestigial sideband

A method of reducing the bandwidth and retaining some of the carrier produces a spectrum whereby the whole of one sideband is transmitted together with a reduced carrier and a shaped portion of the other sideband. This method is used for the broadcasting of television programmes. Vestigial sideband transmission is necessary when a reduction of bandwidth is required but the sideband frequency is too close to the carrier for a filter to be used to remove one sideband. This occurs with television because the modulating frequency can be almost zero. In telephony the lowest frequency is 300 Hz.

Frequency modulation

In this modulation system, the modulating signal causes the carrier to increase and decrease from its initial frequency while its amplitude remains constant. This has the advantages that the radiated signal level remains constant, which allows transmitters to be run at a constant power output, and amplitude variations caused by noise are not demodulated as signals.

The amplitude of the modulating signal is carried in the deviation of the carrier from its original frequency while the frequency of the modulating signal is carried in the cycles of the changes in deviation. The amount by which the frequency of the carrier changes with amplitude changes of the modulating signal is determined by the designer. The greater the change in frequency the greater will be the demodulated signal and the better will be the signal-to-noise ratio. However, although there are many advantages to frequency modulation there is the major disadvantage that a larger bandwidth is required than for amplitude modulation.

(The maximum deviation of the carrier for the maximum modulating signal is known as 'rated system deviation'. The BBC broadcast specification for VHF radio is a rated system deviation of 75 kHz and a modulating frequency of 15 kHz.)

Instead of only an upper and lower sideband as in amplitude modulation, frequency modulation, in theory, produces an infinite number of sidebands.

Which sidebands are important depends upon how much power each sideband possesses. This varies depending upon the modulation index:

$$\text{Modulation Index} = \frac{\text{frequency deviation}}{\text{modulating frequency}}$$

In practice the modulation index can be used to tabulate the value of the power in the sidebands from tables of Bessel functions which list the value of the individual sidebands. Power varies in the individual sidebands with a different modulation index. For modulating indexes below 0.5 the sidebands are restricted to first-order sidebands and the bandwidth is the same as amplitude modulation.

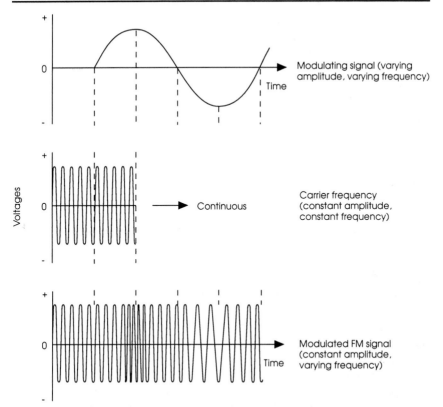

Figure 3.7 *The modulated FM signal varies in frequency depending upon the amplitude of the modulating signal. When there is no modulating signal the carrier frequency is transmitted. The frequency rises with positive modulating amplitudes and falls with negative modulating amplitudes. The frequency deviation is determined by the designer of the equipment.*

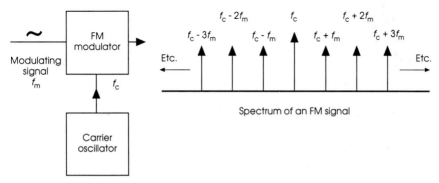

Figure 3.8 *The bandwidth of an FM signal contains many sidebands. Which sidebands need to be transmitted, in order to avoid distortion, is determined by the power in the sideband. This is determined by the modulation index.*

The FM carrier is deviated so that the frequency rises above and falls below the unmodulated frequency. The total frequency variation is known as the 'frequency swing'. The bandwidth, as previously explained, is greater than this and is determined by the modulation index.

An empirical formula, however, allows the bandwidth to be calculated where sidebands of less than 10% of the unmodulated carrier level are ignored. This formula is:

Bandwidth = $2(F_{dev} + F_m)$

The BBC VHF radio bandwidth from the previous specification would be:

Bandwidth = 2(75 kHz + 15 kHz)
Bandwidth = 180 kHz

If double sideband AM modulation was used the bandwidth would be:

2 × modulating frequency
2 × 15 kHz = 30 kHz

If the necessary sidebands of an FM signal are not transmitted, the signal will be distorted.

Advantages of frequency modulation

The problems associated with frequency modulation are related to the larger bandwidth required for transmission and the more complex circuitry required in receivers. However, with integrated circuits this latter problem no longer exists. There are, however, distinct advantages if frequency modulation can be used. These include:

The higher efficiency which can be obtained from transmitters due to the use of class C amplifiers in the RF stages. In addition the individual stages can be operated at maximum efficiency due to the FM signal being a constant amplitude.

As previously explained, selective fading is not a problem with frequency modulation as the amplitude is constant and the system does not respond to any amplitude variations.

A feature of FM reception is the ability of the receiver to select only the more powerful signal if two signals arrive at the aerial at, or close to, the same frequency. The ability of a receiver to perform this function is known as the capture ratio and is given in dB. The dB value gives the level of a received signal below the wanted signal to which the receiver will not respond.

The FM signal provides a better signal-to-noise ratio on the receiver output providing sufficient deviation is used. Unfortunately narrow band FM systems do not have this advantage as the deviation is small in order to reduce bandwidth.

When designing systems, frequency modulation allows both a better dynamic range and linearity to be provided in the equipment. This means that on multichannel systems crosstalk is reduced and the ability to transmit both the smallest and largest signals is improved when compared to AM modulation.

Noise

Whenever a signal is transmitted by cable, radio or through any piece of equipment, noise in the form of unwanted signals is added to the waveform as amplitude variations. If the wanted signal is also in the form of amplitude variations, then the noise is inseparable from the signal. If the noise is of the same frequency as the signal it is not possible to filter it without affecting the signal.

All components produce noise and this increases with heat. When signals are very small, great care has to be taken with the design of the first amplifying stages in order that the noise produced by the stage does not swamp the original signal. Sometimes the amplifiers are enclosed in supercooled conditions, as was the situation at the Goonhilly earth station for the reception of the first satellite signals. How the noise affects a signal depends upon the type of signal. Frequency modulated signals and digital signals are less affected by noise as their form provides some immunity when recovering the wanted signal (see later). Noise, however, is still added and eventually determines when a signal ceases to be usable.

Sources of noise

Noise can come from many sources. It can be natural, man-made, component noise, interfering signals from other radio sources and, in multi-channel systems, intermodulation effects can be caused by non-linearity in the transmission system.

Natural noise includes:

1. Cosmic noise which originates in outer space and can affect radio links.
2. Electrical storms causing lightning.
3. Static due to storms in the ionosphere which cause the earth's magnetic field to fluctuate. These storms are caused by the sun and are most troublesome when sun spot activity is high. Radio interference occurs a few days after large sun spots occur as particles from the sun enter the ionosphere on the solar wind.

Man-made noise includes:

1. Car ignition systems.
2. Electrical switching.
3. Electrical equipment which radiates electromagnetic waves (motors and generators with brushes, televisions, computers, invertors, radios, etc., and any electrical sparking).

Equipment noise includes:

1. Hum from power supplies.
2. Random movement of atoms in components which is determined by heat.
3. Shot noise caused by active devices such as transistors and valves. It is caused by the varying velocity of the electrons through the device when potentials are applied at the terminals.
4. Partition noise which occurs in multi-electrode devices due to the current being divided between the electrodes.

5. Flicker noise, a low frequency noise which is not fully understood. It occurs mainly below 1 kHz and decreases with an increase in frequency.

6. Microphonic noise caused by mechanical vibration of parts of the circuit. Valves, camera tubes, microphones, sensors, etc., are affected and equipment may have to be mounted on shock absorbers.

Multi-channel systems

Serious noise problems occur in multi-channel systems where many frequencies are passing through amplifiers. Any non-linearity in the amplifier causes intermodulation between the frequencies. This produces new frequencies which may fall within the bandwidth of other channels. The higher the power in the amplifiers and the more channels that are transmitted the greater can become the problem. It is often a limiting factor on cable television systems and carriers have to be carefully chosen to ensure the products of intermodulation fall in areas which give the least problems. Amplifier circuitry has to be devised to reduce the problem and push-pull arrangements are used in order to eliminate second harmonic distortion.

When channels are adjacent to each other, interference between the channels can occur when filters and the selectivity of receivers are inefficient and allow signals from adjacent channels to be received.

These same problems have also to be solved in multi-channel telephony when frequency division multiplexing (FDM) is used and thousands of telephone conversations are simultaneously transmitted down a single cable or radio link with each telephone conversation having been modulated several times and finally occupying a unique band of frequencies.

Crosstalk

When cables carrying different signals are laid adjacent to each other, coupling can occur between the cables due to electromagnetic induction. Electrical signals in one pair create a magnetic field which cuts the adjacent pair and induces signal voltages. This is known as crosstalk and can be reduced by screening, twisting pairs and using balanced circuits which allow induced voltages to cancel each other out at the receiving end. By balancing the circuits to earth, the induced voltage will be in the same direction and equal in each wire of the pair. These will cancel in a balanced transformer without affecting the required signal which, because it is a loop, is flowing in opposite directions in each wire.

Multiple hop radio links

When multiple hop links are used to cover a long distance it is essential that receivers do not pick up signals of the same frequency from other than the desired transmitter. This can be eliminated by careful route planning to ensure transmitters are not in line and by changing the frequency of each stage. Any overspill from a previous transmitter is not selected by the receiver.

When groups of frequencies are reused in a radio system it is essential to ensure that the power of the transmitters and the distances apart are carefully engineered

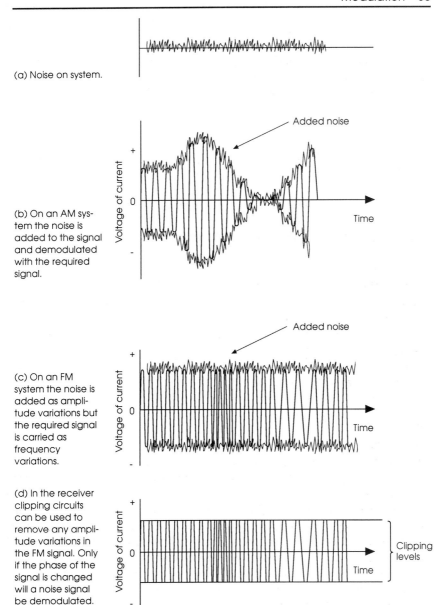

(a) Noise on system.

(b) On an AM system the noise is added to the signal and demodulated with the required signal.

Added noise

Voltage of current

Time

(c) On an FM system the noise is added as amplitude variations but the required signal is carried as frequency variations.

Added noise

Voltage of current

Time

(d) In the receiver clipping circuits can be used to remove any amplitude variations in the FM signal. Only if the phase of the signal is changed will a noise signal be demodulated.

Voltage of current

Time

Clipping levels

Figure 3.9

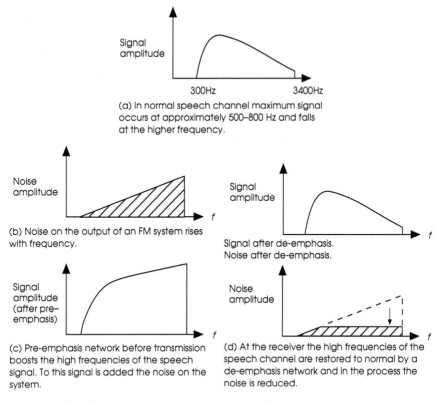

(a) In normal speech channel maximum signal occurs at approximately 500–800 Hz and falls at the higher frequency.

(b) Noise on the output of an FM system rises with frequency.

Signal after de-emphasis.
Noise after de-emphasis.

(c) Pre-emphasis network before transmission boosts the high frequencies of the speech signal. To this signal is added the noise on the system.

(d) At the receiver the high frequencies of the speech channel are restored to normal by a de-emphasis network and in the process the noise is reduced.

Figure 3.10   *Pre-emphasis in FM system.*

to ensure radio waves are not received in areas where receivers could select their signals.

The amount of noise from all the mentioned sources becomes the limiting factor when the acceptability of a signal is considered. The signal at all times must maintain a level above the noise which ensures it is usable without fatigue and annoyance to the listener or viewer. The necessary level of separation of noise and signal is different for each type of signal and the methods of coding and modulation for transmission are often chosen to improve this parameter. The ratio of signal to noise is a most important figure in any communication system and all system engineering revolves around obtaining an acceptable figure.

Noise on a modulated waveform

The required signal from an amplitude modulated waveform is in the form of varying voltage amplitudes. When noise is added from any source, it is also in the form of varying voltage amplitudes. In the receiver the demodulation system is unable to differentiate between the two signals and the noise is demodulated as a signal.

A frequency modulated signal carries no information in the amplitude variations of the waveform as the signal is in the changing frequency of the carrier. Any noise is still added as amplitude variations and the advantage of FM is that clipper circuits can be used to remove the amplitude variations without affecting the required signal. If, however, the noise causes the phase of the FM signal to vary then this signal will be demodulated as noise. The level of the noise signal produced is directly proportional to frequency and is, therefore, highest at the higher frequencies. An improvement in signal-to-noise ratio is obtained by an increase in rated system deviation and pre-emphasis of the higher frequencies before transmission (see Figure 3.10).

Pre-emphasis is used in FM systems to boost the top frequencies before modulation at the transmitter and de-emphasis of equal and opposite amount is used at the receiver to restore the signal to its original levels. However, in the process the noise at the higher frequencies is reduced. When used as a speech channel, the human voice produces most of its power at the lower frequencies. The bandwidth of commercial speech, 300-3400 Hz, contains most of the power at approximately 600 Hz (a man's voice is lower in frequency than a woman's) but frequencies around 1800 Hz, although small in intensity, are also necessary for good articulation. As has been previously seen the noise in an FM system rises with an increase in frequency. Pre-emphasis, therefore, improves the signal-to-noise ratio at the higher frequencies where signals are naturally smaller.

Decibels

The design engineer and maintenance technician spend a considerable amount of their time measuring signal levels and calculating gains and losses in a system. It is essential in all communication systems that the correct levels are maintained at all times to ensure the signal is high enough for a satisfactory signal-to-noise ratio but not so high that amplifiers are overloaded and distortion and intermodulation effects are introduced into the system.

Communication systems comprise amplifiers which multiply the level of the signal on the input and equalisers, attenuators and system losses, such as cables, which divide the signal on the input. To calculate the overall gain or loss using division and multiplication on a complex system comprising many pieces of equipment and transmission medium becomes a tiresome and difficult task. A system whereby gains are added and losses are subtracted simplifies the task enormously. The necessary gain or loss required to bring the system to its correct level can be quickly calculated and implemented. The gains of amplifiers and the losses of attenuators can be calibrated in units which allow the necessary adjustments to be made quickly and accurately with the minimum of mathematics.

The solution

Before the calculator became an everyday item, multiplication of numbers was performed by finding the logorithms of the individual numbers and adding them. Division was performed by finding the logorithms of the numbers and subtracting them. If we, therefore, convert the gains and losses into logarithms we have exactly the system that is required to simplify the mathematics of system engineering.

Bel and decibel

Communication systems are designed to pass power between two points: from the system input to the system output. The logarithmic unit of the ratio of the power input and the power output is known as the bel and is given as:

$$\text{Bel} = \log\left(\frac{\text{power output}}{\text{power input}}\right)$$

This is the definition for a power gain in bels.

The unit is too large for general use if whole numbers are desired. A smaller unit, known as the decibel, is, therefore, the most commonly used:

$$\text{Decibel} = 10\log\left(\frac{\text{power output}}{\text{power input}}\right)$$

The decibel is a logarithmic ratio and has no units.

If we wish to work in voltages or currents we use the formula:

$$\text{Decibel} = 20\log\left(\frac{\text{voltage out}}{\text{voltage in}}\right) \quad \text{or } 20\log\left(\frac{\text{current out}}{\text{current in}}\right)$$

This is the voltage or current gain in decibels.

These formulae are derived from the fact that:

$$\text{Power} = \frac{V^2}{R} \text{ or } I^2R$$

Substituting into the formula:

$$\text{Decibel} = 10\log\left(\frac{V^2_{out}/R}{V^2_{in}/R}\right)$$

(input and output impedances (R) must be equal.

$$\text{Decibel} = 10\log\left(\frac{V^2_{out}}{V^2_{in}}\right) = 10\log\left(\frac{V_{out}}{V_{in}}\right)^2$$

$$\text{Decibel} = 20\log\left(\frac{V_{out}}{V_{in}}\right)$$

A similar proof is used for current using I^2R for power.

Calculating system gains and losses

If all the individual gains and losses in a system are quoted in decibels, the gains can be added and the losses subtracted in order to find the overall gain or loss of the system. It becomes a simple exercise to decide how many decibels gain an amplifier must have or how many decibels loss an attenuator must have to produce the required overall level of the system.

Signal-to-noise ratio

This is a most important figure for any system or individual item of equipment and it determines the quality of the signal:

$$\text{It is quoted in decibels} = 10 \log \left(\frac{\text{signal power}}{\text{noise power}} \right)$$

$$\text{or decibels} = 20 \log \left(\frac{\text{signal voltage}}{\text{noise voltage}} \right)$$

In a communication system, the necessary signal-to-noise ratio that is required for adequate signal quality is determined by the type of signal being transmitted.

Typical values include:
Excellent television pictures 45 dB
Marginal television pictures 28 dB
Digital transmission on cable network better than 22 dB
High quality audio 80 dB
Telephone conversation over public network 35 dB
Music circuit for broadcasting 60 dB
Private land mobile telephone systems 10 dB, Etacs cellular 17 dB minimum
Ship-to-shore radio telephone systems 20 dB

Noise occurs throughout the spectrum and it must be remembered that the greater the bandwidth of the communication channel the more noise is introduced into the system.

Reference levels

As the decibel is only a logarithmic ratio, it must be referred to a reference in order to know the voltage or power a decibel value represents.

Many engineers from different disciplines are able to use decibels for measuring by each one using a different reference. The reference is made 0 dB and then different values of dB relate directly to specific values. Engineers specialising in acoustics, video, audio, radio and transmission each have their own reference. Typical references are:

Video: 1 volt across 75 ohm
Transmission: 1 milliwatt in 600 ohms
Radio: 1 millivolt across 75 ohms
Acoustics: a pressure of 20 micropascals (threshold of hearing)

When the dB value directly relates to a reference it is written to show the reference (dBmV, etc). Table 3.1 shows the dB value directly related to voltage for a reference of 1 mV across 75 ohms being equal to 0 dBmV

Calculations using decibels

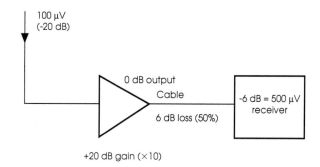

+20 dB gain (×10)

EXAMPLE 1 *An aerial receives 100 μV of signal and is fed into a +20 dB amplifier. The cable connecting the amplifier to the receiver has a 6 dB loss. Use Table 3.1 to find the signal voltage at the receiver.*

EXAMPLE 2 *There is a series of amplifiers and losses on a transmission line as shown in the diagram. The imput signal is 1 V and it is required to have 1 V on the output. Is A a loss or an amplifier?*

The overall grain of system is 0 dB.
The gain of the amplifiers = + 43 dB
The Losses = – 31 dB
———————
Difference = + 12 dB

A must be an attenuator (loss) to make the system unity gain (0 dB). Its value is –12 dB.

Table 3.1 *dBmV/microvolt chart*

dBmV	μV	dBmV	μV	dBmV	μV
−40	10	0	1000	40	100 000
−39	11	1	1100	41	110 000
−38	13	2	1300	42	130 000
−37	14	3	1400	43	140 000
−36	16	4	1600	44	160 000
−35	18	5	1800	45	180 000
−34	20	6	2000	46	200 000
−33	22	7	2200	47	220 000
−32	25	8	2500	48	250 000
−31	28	9	2800	49	280 000
−30	32	10	3200	50	320 000
−29	36	11	3600	51	360 000
−28	40	12	4000	52	400 000
−27	45	13	4500	53	450 000
−26	50	14	5000	54	500 000
−25	56	15	5600	55	560 000
−24	63	16	6300	56	630 000
−23	70	17	7000	57	700 000
−22	80	18	8000	58	800 000
−21	90	19	9000	59	900 000
−20	100	20	10 000	60	1.0 volt
−19	110	21	11 000	61	1.1
−18	130	22	13 000	62	1.3
−17	140	23	14 000	63	1.4
−16	160	24	16 000	64	1.6
−15	180	25	18 000	65	1.8
−14	200	26	20 000	66	2.0
−13	220	27	22 000	67	2.2
−12	250	28	25 000	68	2.5
−11	280	29	28 000	69	2.8
−10	320	30	32 000	70	3.2
− 9	360	31	36 000	71	3.6
− 8	400	32	40 000	72	4.0
− 7	450	33	45 000	73	4.5
− 6	500	34	50 000	74	5.0
− 5	560	35	56 000	75	5.6
− 4	630	36	63 000	76	6.3
− 3	700	37	70 000	77	7.0
− 2	800	38	80 000	78	8.0
− 1	900	39	90 000	79	9.0
− 0	1000	40	100 000	80	10.0

0 dBmV = 1000 μV across 75 ohms (1mV across 75 ohms)

4 Amplifiers

The design of amplifiers varies considerably depending upon the purpose for which they are used. They basically fall into two categories:

1. Voltage amplifiers which are designed to produce signal voltage gain between the input and the output of the amplifier.
2. Power amplifiers which are designed to drive such devices as a loudspeaker, a transmitting aerial or transmission line and mechanical items. To drive such devices requires power which means both voltage and current must be supplied by the amplifier.

The design of both types of amplifiers must take into consideration:

1. The voltage gain or power output requirement.
2. The bandwidth of the signal being amplified.
3. The amplitude of the input and output signals.
4. The noise and distortion introduced by the amplifier.

Amplifiers are required for:

1. Audio (signals in the region of 20 Hz to 20 kHz.)
2. Video (signals in the region of 0 to 6 MHz)
3. Radio frequencies (signals ranging from kHz to MHz)
4. Ultra high frequencies (signals exceeding 300 MHz)

Amplifiers which are designed to cover a wide range of frequencies are known as wide-band amplifiers and must provide linear performance throughout the band.

In radio engineering amplifiers are designed to cover a selective range of frequencies occupied by a single channel. These are known as tuned amplifiers.

Each type of amplifier requires different construction techniques depending upon the operating frequencies, the bandwidth of the signals and the input and output signal levels. At high frequencies stray capacitance within the circuit layout and components becomes important as it can act as a coupling component between different parts of the circuit which can cause oscillations, short circuits to high frequencies or unwanted negative feedback. The stray capacitance has a reactance which is inversely proportional to frequency:

$$X_c = \frac{1}{2\pi f C} \text{ ohms}$$

At low frequencies the capacitive reactance is very high and has very little or no effect on the operation of the circuit. At high frequencies the reactance is very low and becomes a limiting factor for the amplification and transmission of high frequencies.

As previously described, all components, whether passive or active, introduce noise into a system. Any non-linearity in an amplifier introduces both linearity distortion, which changes the ratio of the various amplitudes of the signal between the input and output of the amplifier, and intermodulation between the different frequencies present. The higher the voltage gain or power output of the amplifier the more likely this effect occurs. It is especially important that the introduction of noise is minimal when the signal is very small otherwise amplifier noise can swamp the wanted signal. Pre-amps, which are the first stage of amplification of many systems, are designed with active amplifying devices which introduce minimum noise and are often special low noise transistors or field effect transistors (FETs) which have a lower noise characteristic than a normal transistor. The circuit is also often screened with metal to protect it from the radiation of oscillators, transformers and other radiated signals which can be coupled into the signal chain. The power supply, which must also be smooth to prevent hum, can be an important channel for introducing noise into the system from the mains supply or coupling from other circuits. Decoupling capacitors on each amplifying stage are often used to minimise the problems. The pre-amp is placed as close to the source of the signal as possible in order that pickup is not experienced on the connection between the signal device (tape head, camera tube, etc.) and the amplifier.

Power amplifiers

In very high power amplifiers, especially in transmitters where the power can be many kilowatts, transistors are unable to provide sufficient output power and valves are still used. The valves are triodes or tetrodes although the initial amplifying stages may still use transistors. Small power amplifiers can, however, be fully transistorised and single integrated circuits for both amplification and power output can be used to drive loudspeakers in radios and televisions.

Maximum power transfer theorem

All signal sources have an internal impedance and if power is to be delivered into a load the relationship between the internal impedance and the load impedance is critical. In order for maximum power to be transferred to the load the impedance of the load must equal the internal impedance of the power source (see Figure 4.1(a)). If the

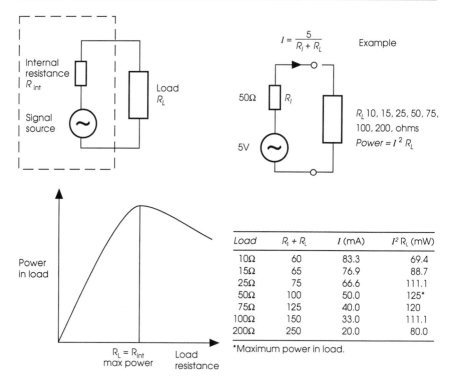

Figure 4.1(a)   *Maximum power is transferred to load when the load resistance equals the internal resistance of the source.*

load and the internal impedance are different a matching transformer can be used to make the load appear to be the correct impedance to the source (Figure 4.1(b)).

The same principle applies to a transmitter feeding an aerial or an amplifier feeding a transmission cable. If the source and the load are not matched there is a loss of power transfer (see the section on transmission lines). There is also a further problem on transmission lines feeding aerials and networks when mismatch occurs at the end of the line as standing waves and reflections result. This not only leads to a loss of power but also results in signals being reflected from the end of the cable to the output of the amplifier. It must be remembered that power can only be absorbed in a resistive load which must appear at the end of any transmission line in order to avoid problems.

Classification of amplifiers

Amplifiers are classified according to the method used for biasing the stages. In a common emitter transistor amplifying circuit (Figure 4.2) the load resistor is connected between the collector and the supply voltage. A signal is supplied to the base which causes a current variation in the base–emitter circuit (I_b) which in turn causes a

Figure 4.1(b) *Matching transformer.*

When the load and source are not equal a matching transformer can be used in order to obtain maximum power transfer from source to the load.

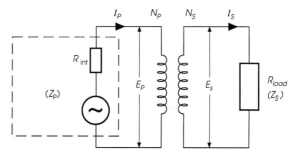

Power in primary = power in secondary

$$E_p I_p = E_s I_s$$

$$\frac{N_p}{N_s} = \frac{E_p}{E_s} = \frac{I_s}{I_p}$$

$$\frac{N_p}{N_s} \times \frac{N_p}{N_s} = \frac{E_p}{E_s} \times \frac{I_s}{I_p}$$

$$\left(\frac{N_p}{N_s}\right)^2 = \frac{E_p}{I_p} \times \frac{I_s}{E_s} = \frac{Z_p}{Z_s}$$

N_p = number of turns on the primary

N_s = number of turns on the secondary

n = turns ratio

$$n = \frac{N_p}{N_s} = \sqrt{\frac{Z_p}{Z_s}} \quad \text{where } n > 1$$
$$Z_p > Z_s$$

When using a matching transformer the higher impedance is always connected to the winding with the greatest number of turns.

EXAMPLE

The primary impedance is 1000 Ω. The secondary circuit impedance is 10 Ω. The secondary windings have 200 turns. How many turns need to be on the primary to match impedances?

$$n = \sqrt{\frac{1000}{10}}$$

$$n = \sqrt{100} = 10$$

$$\frac{N_p}{N_s} = 10$$

$$N_p = 10 \times 200$$

$$N_p = 2000 \text{ turns}$$

much higher current variation in the collector circuit (I_c). The current variations across the load resistor in the collector circuit produce a signal voltage ($I_c R_L$) which is far higher than the signal voltage applied to the base. The amplifier stage, therefore, produces a voltage gain if the output signal is taken from across the load resistor. When there is no signal applied to the base the current flowing in the collector circuit depends upon the bias applied to the base.

Class A operation   (Figure 4.2)

In this circuit, the bias is chosen so that with no input signal the collector current is sufficient to cause a voltage drop across the load which is equal to half the supply voltage. The collector voltage is at a value which is then at the middle value of the supply voltage. When a signal is applied to the base, the collector is able to swing linearly in both a maximum and a minimum direction without distortion. This type of circuit is used for voltage amplification and for single transistor output stages where linearity is essential. Class A circuits produce a large signal voltage output but are inefficient for use as a power amplifier where it is generally less than 25% efficient. It must be remembered that it is the supply power which is being converted to signal power in a power amplifier.

Class B operation   (Figure 4.3)

In class B operation the bias is adjusted so that with no input signal there is no collector current. The transistor is biased to cut off. When a signal is applied, collector current only flows for half a cycle and the other half is cut off. In order to obtain a complete output signal a circuit must be arranged so that two output transistors are used and each transistor is made to handle half the signal. On the output the two halves must be combined. This means that each transistor operates in turn and the input signal is split in such a way that each transistor amplifies either the positive or the negative parts of the input signal. The circuits are known as push–pull amplifiers and either transformers are used for splitting the phase of the input signal and combining the output signals or circuits are constructed with complementary circuits, using matched NPN and PNP transistors. These latter circuits lend themselves to being made into integrated circuits.

The advantages of push-pull circuits are:

1. There is very little current drawn from the supply when there is no input signal.
2. Each transistor only works for half a cycle and, therefore, its rating can be five times the normal power rating of the transistor.
3. Because the current is symmetrical, any induced interference is cancelled and does not appear in the load.
4. Up to 78.5% of the DC power supplied can be converted into output signal power.

It is essential that the two halves of the circuit match otherwise distortion occurs. It is also important that each transistor is slightly forward biased otherwise crossover distortion occurs when the two output signals are combined due to the non-linearity of a transistor when it begins to conduct.

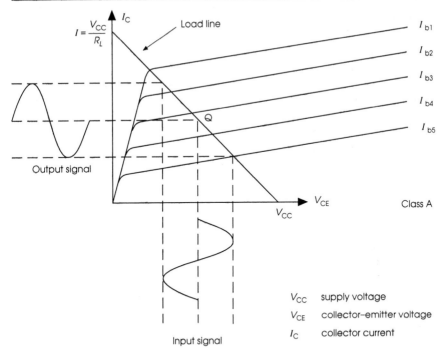

The bias is chosen so that the input signal causes a linear output. When there is no input signal a current I_c flows which is determined by the quiescent bias point Q.

The load line is constructed by two calculations. When maximum current flows in collector, V_{CE} is zero and current is determined by R_L:

$$I = \frac{V_{cc}}{R_L}$$

When no collector current flows $V_{CE} = V_{CC}$. These conditions are the two extremes. When an input signal is applied the dynamic operation of the circuit is determined by the load line.

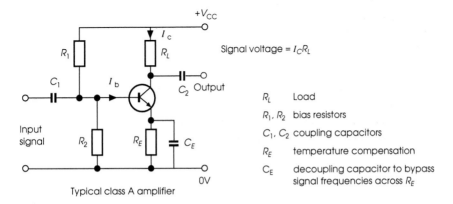

Signal voltage $= I_C R_L$

R_L Load

R_1, R_2 bias resistors

C_1, C_2 coupling capacitors

R_E temperature compensation

C_E decoupling capacitor to bypass signal frequencies across R_E

Typical class A amplifier

Figure 4.2   *Class A amplifier (common emitter amplifier).*

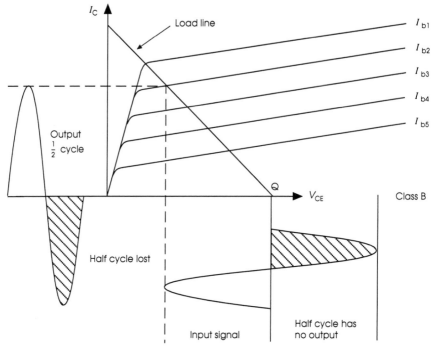

The bias point is chosen as the cut-off point on the load line. Only one half cycle is amplified. Two transistors are used in a push–pull circuit in order to obtain the complete output signal.

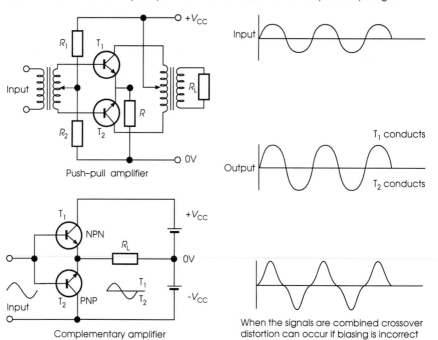

Figure 4.3   *Class B amplifier*

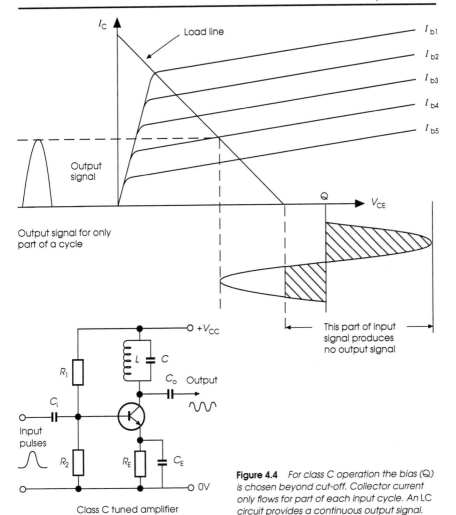

Figure 4.4 *For class C operation the bias (Q) is chosen beyond cut-off. Collector current only flows for part of each input cycle. An LC circuit provides a continuous output signal.*

Class C tuned amplifier

Class C operation (Figure 4.4)

In class C operation the bias is increased until it is at a value beyond cut-off. The output signal only occurs for part of each input cycle. Obviously such amplifiers are of little use for amplifying signals containing linear information as the distortion cannot be rectified. Class C operation is used with tuned circuits which naturally oscillate and restore any sinusoidal waveform. The inductor and capacitor tuned oscillating circuit, which is the transistor load, has a natural resonant frequency and the transistor only provides the necessary power which is sufficient to overcome any circuit losses. These circuits can approach 100% efficiency.

It will be seen in Chapter 5 that class C amplifiers are used extensively in transmitters to amplify the carrier frequency and provide efficient power amplifiers.

They are also used in frequency multipliers where the tuned circuit is tuned to a harmonic of the input signal. Class C operation is also used in oscillators.

Feedback

Feedback involves part of the output signal of an amplifier being combined with the input signal. The phase of the feedback signal can cause the feedback signal to:

1. Add to the input signal.
2. Subtract from the input signal.

When the input and output signals add it is known as positive feedback and this provides the method for designing oscillators. When negative feedback is applied the portion fed back from the output is 180 degrees out of phase with the input signal and the signals subtract from each other. This considerably reduces the gain of the amplifier stage which may initially appear to be a disadvantage. However, there are so many advantages that all high quality amplifiers incorporate negative feedback in their design. The effects of negative feedback are:

1. The gain of the amplifier is reduced.
2. The gain of the amplifier is made more stable.
3. The bandwidth of the amplifier is increased.
4. Distortion is reduced.
5. Noise is reduced.
6. Input and output impedances are changed.

In order to obtain the same gain with negative feedback as with an amplifier without negative feedback it may be necessary to use two or more stages in cascade. This is obviously an additional expense and complication in the design of amplifiers. However, modern integrated circuits allow high gain operational amplifiers to be produced as standard items. A designer need only use standard operational amplifiers and modify their characteristics with negative feedback by the addition of a few external components to obtain an amplifier of the required specification.

The gain of an amplifier with feedback is given by:

$$G_v = \frac{A_v}{1 - \beta A_v}$$

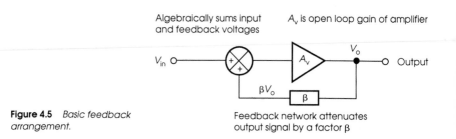

Figure 4.5   *Basic feedback arrangement.*

where G_v = gain with feedback
         β = portion of output signal fed back to input
         A_v = gain of amplifier without feedback

If β is positive the gain is increased and oscillation will build up (positive feedback).
   If β is negative

$$G_v \fallingdotseq \frac{A_v}{1 + \beta A_v} \quad \text{and the gain is reduced}$$

If $\beta A_v > 1$ then $G_v = \dfrac{1}{\beta}$

and the gain is controlled almost solely by the amount of feedback and not by the actual gain of the amplifier. This means that if components age or change their value with temperature variations and transistor gains change with time, the gain of the stage still remains fairly constant. This is important, especially when equipment has a long life expectation or operates in hostile conditions or where regular maintenance is either not possible or desirable. It is also important where consistency of performance is necessary.

Bandwidth

The bandwidth of an amplifier is restricted by the coupling capacitors between stages at the lower frequencies and by shunt or stray capacitance across the components and circuitry at the higher frequencies. In the first instance the capacitance is a series impedance which attenuates the signal at the lower frequencies and in the second instance it is a low impedance across the components which short circuits the higher frequencies (X_c is inversely proportional to frequency $X_c = 1/2\pi f_c$).

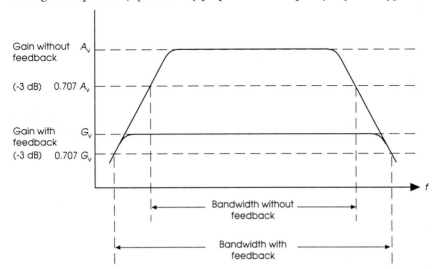

Figure 4.6 *The bandwidth of an amplifier is increased with negative feedback.*

Bandwidth is defined as the frequency response between the points which are at half power or 3 dB below the maximum (0.707 max) of either A_v or G_v. It can be seen from Figure 4.6 that the application of negative feedback increases the effective bandwidth by the same factor as the gain is reduced.

Noise and distortion

Noise and distortion are reduced by a factor which depends upon the amount of negative feedback applied:

$$\text{Output of noise and distortion} = \frac{D_o}{1 + \beta A_v} + \frac{N_o}{1 + \beta A_v}$$

where N_o = noise without feedback
D_o = distortion without feedback
β = proportion of output signal fed back to input
A_v = open loop gain of amplifier without feedback

Example
The open loop gain of an amplifier is 250 and the proportion of output fed back as negative feedback is 0.2. If the distortion without feedback is 5% and noise is 0.1 mV find the distortion and noise on the output of the amplifier with negative feedback.

$$\text{Output} = \frac{5}{1 + 250 \times 0.2} + \frac{0.1}{1 + 250 \times 0.2}$$

$$\text{Output} = \frac{5}{51} + \frac{0.1}{51}$$

Distortion on output with feedback = 0.098%
Noise on output with feedback   = 0.002 mV

Feedback arrangements

There are alternatives for both the input and output feedback arrangements in order to combine the feedback and input signals. These involve either a series or shunt arrangement on the input or output and the feedback signal can be derived from either the output current or voltage.

If a series arrangement is used the impedance is increased whereas a shunt arrangement decreases the impedance. In order to derive the feedback from the current a series resistor is incorporated in the output circuit and a voltage directly proportional to the current is obtained as the feedback signal. For voltage feedback a high impedance resistive network is placed across the output and the correct proportion of the voltage is obtained by the choice of the ratio of the resistor values.

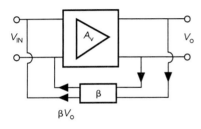

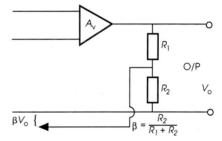

(a) Shunt voltage feedback. Feedback applied in parallel with input signal

(b) Feedback voltage from across R_2:β. This is a practical circuit for shunt voltage feedback. Feedback voltage = βV_o

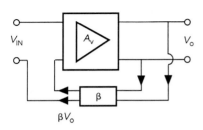

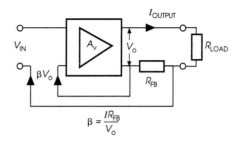

(c) Series voltage feedback. Feedback applied in series with input signal

(d) Series current feedback. Feedback applied in series with input signal.
Feedback voltage = IR_{FB}. Feedback voltage is proportional to load current (R_{FB} must be small compared to R_{LOAD})

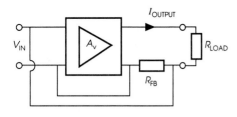

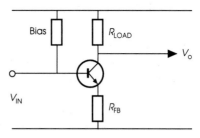

(e) Shunt current feedback. Feedback applied in parallel with input signal

(f) A practical current feedback circuit. An undecoupled emitter resistance provides current feedback

Figure 4.7 *Negative feedback arrangements.*

A table can be made to show the effect of the different arrangements:

Table 4.1

Feedback arrangement	Comparison with amplifier without feedback	
	Input impedance	Output impedance
1. Series current	increases	increases
2. Shunt current	decreases	increases
3. Series voltage	increases	decreases
4. Shunt voltage	decreases	decreases

Oscillators

Highly accurate oscillators are essential for the orderly control of the radio spectrum. It is the radio frequency oscillators which produce the carriers for all the modulating systems which are transmitted by any type of radio device whether it is a broadcasting transmitter, mobile radio or portable phone.

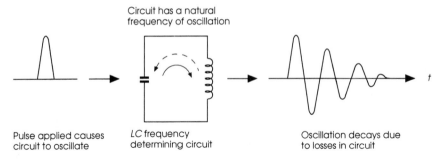

Circuit has a natural
frequency of oscillation

Pulse applied causes   *LC* frequency   Oscillation decays due
circuit to oscillate   determining circuit   to losses in circuit

In order to maintain oscillation pulses must be applied at the correct frequency and phase which are sufficient to replace lost energy due to losses in the circuit. The oscillation is maintained by a transfer of energy between the inductor and capacitor. Assume that *C* is fully charged. It discharges through *L* and the current flowing causes a magnetic field around the inductor. When *C* has discharged, the current stops and the magnetic field collapses causing an induced voltage in *L*. This causes a reverse current and *C* again charges. The cycle is again repeated. Resistance in the circuit causes energy loss.

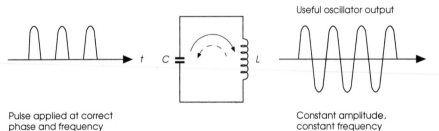

Useful oscillator output

Pulse applied at correct   Constant amplitude,
phase and frequency   constant frequency

Figure 4.8   *Maintenance of oscillation waveform in an LC tuned circuit*

When a single pulse of energy is given to any unit capable of producing a sinusoidal waveform (sine wave) the output gradually dies away due to the losses in the system. In order for a system to maintain oscillation, positive feedback is required whereby the output signal is applied to the input at the correct time, phase and amplitude in order to overcome losses and maintain oscillation (see Figure 4.8).

For a circuit to oscillate at a single frequency there must be built into the circuit a frequency determining unit. This circuit controls the frequency of the feedback signal and, therefore, the output frequency.

Oscillators are required not only in transmitters but also in receivers to heterodyne with the incoming signal and produce the intermediate frequency required in superhet receivers (see radio receivers, Chapter 6). Scanning waveforms in television sets and cameras rely on oscillators producing accurate sawtooth waveforms to deflect the electronic beams across the camera tubes and cathode ray viewing tubes and all parts of a television broadcast system are synchronised by a sync pulse generator which produces all the necessary timing pulses. Computers rely on the generation of pulse producing circuits such as monostables, bistables and astables for their operation and all the circuitry relies on the production of accurate clock pulses to activate the circuits and functions at the correct time. All digital circuitry in transmission systems and digital telephone exchanges relies on the accurate generation of timing signals from pulse oscillators for coding, decoding and switching. Pulse oscillators are used in radar systems and pulse and analogue oscillators are used in test equipment.

Types of oscillators

Relaxation oscillators

The basic astable circuit consists of two transistors whose collectors and bases are cross coupled with a capacitor and resistor timing circuit. At any single time one transistor is switched on and conducting while the other is switched off. After a period of time, determined by the values of the CR network, the transistors are switched to their opposite state. The output on the collector of either transistor is, therefore, a free running square wave whose mark–space ratio (period on and period off) is determined by the CR values in the individual transistor circuits (Figure 4.9).

The circuit can be modified so that one transistor is biased and only switches when a pulse is fed to its base which is sufficient to overcome the bias and cause it to switch on. This is a monostable circuit and one cycle of pulses is produced for each input pulse. Such circuits are used for retiming the duration of input pulses and accurately produce pulses of the required duration (Figure 4.10).

Biasing both transistors produces a bistable circuit and this requires two input pulses to produce one cycle of output square wave pulses. The circuit divides the input pulses by two and forms the basis of frequency division for clocks and timing circuits. Combinations of these circuits, with feedback arrangements where necessary, can produce any division required. A further important use of these circuits is as a memory in computers as they will change state with each pulse and can, therefore, be used as a 1 bit memory (Figure 4.11).

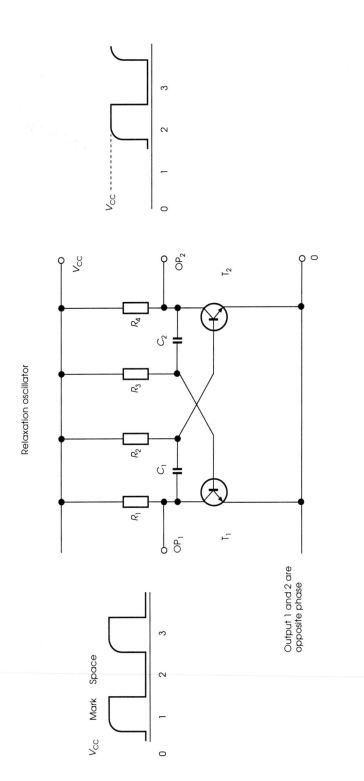

Relaxation oscillator

Output 1 and 2 are opposite phase

Figure 4.9 Frequency of oscillation is governed by the time constants $R_2 C_1$ and $R_3 C_2$. Long time constants provide low frequencies and short time constants high frequencies. If the time constants are equal the mark–space ratio of the pulse outputs will also be equal. Different time constants allow different duration pulses to be generated.

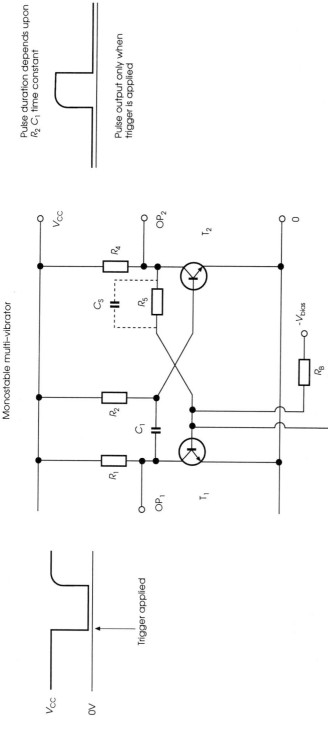

Figure 4.10 *The basic astable circuit is modified to bias T₁ off. T₁ only conducts when a trigger pulse is applied which is sufficient to overcome the bias. The pulse duration which appears on OP₂, when a trigger is applied, depends upon the time constant R₂ C₁. A capacitor is often fitted across R₅ (speed-up capacitor) to make the switching action quicker (Cₛ)*

Bistable multivibrator (flip-flop)

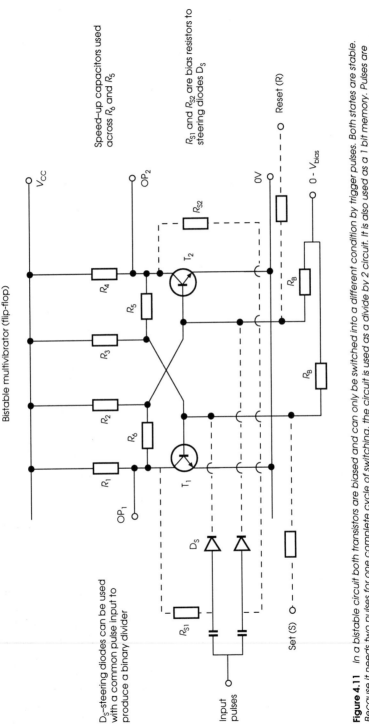

D_S—steering diodes can be used with a common pulse input to produce a binary divider

Speed-up capacitors used across R_6 and R_5

R_{S1} and R_{S2} are bias resistors to steering diodes D_S

Figure 4.11 In a bistable circuit both transistors are biased and can only be switched into a different condition by trigger pulses. Both states are stable. Because it needs two pulses for one complete cycle of switching, the circuit is used as a divide by 2 circuit. It is also used as a 1 bit memory. Pulses are applied to the R and S inputs to set and reset the memory. An RS bistable memory has a problem when pulses are applied simultaneously to the R and S inputs. The switching state can be indeterminate and, therefore, a more complex circuit is normally used and is known as a JK bistable or flip-flop. When the trigger pulses are applied from a common input, diodes are normally used to steer the pulse to the correct transistor. Depending on the transistor state, the diode will be biased to conduct or will be reversed biased.

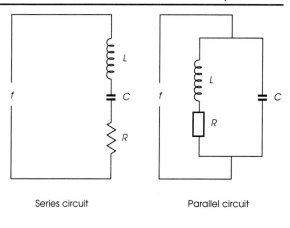

(a) At resonance $X_L = X_C$ and circuits are resistive. A series circuit at resonance is only R as reactances X_L and X_C cancel. A parallel circuit becomes a high impedance at the resonant frequency. The quality of an inductor or capacitor which enables high selectivity is given by its Q factor:

$$Q = \frac{\omega L}{R} \text{ or } \frac{1}{\omega CR} \quad (\omega = 2\pi f)$$

Series circuit

Parallel circuit

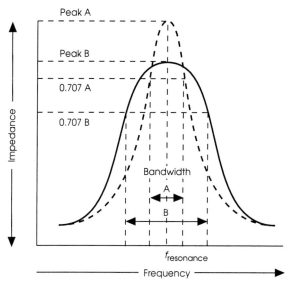

(b) Impedance curves of parallel tuned circuits with different resistance. Peak B has higher resistance and bandwidth is increased. Tuned parallel circuits are used as the loads of RF amplifiers and the bandwidth determines the selectivity of the amplifier. Bandwidth is defined as the frequencies between the points which are 0.707 peak.

Figure 4.12 *Tuned circuits and bandwidth*

Feedback oscillators

These provide sinusoidal waveforms which are necessary for radio carrier waves. The feedback determining unit is placed between the output and the input so that only one frequency is fed back, which is the frequency of oscillation. The oscillators require a 360 degree phase shift in the feedback circuit to make the output signal in phase with the input. This is achieved with a single stage amplifier, which itself produces a 180 degree phase shift between the base and collector, followed by a further 180 degree phase shift introduced by a component such as a transformer. Two inverting amplifying stages in cascade also produce the necessary 360 degree phase shift and feedback can be arranged between the input of one and the output of the

Inductor–capacitor (*LC*) oscillator

Figure 4.13 *The frequency of oscillation is determined by* L_1, C_3

$$f = \frac{1}{2\pi \sqrt{L_1\, C_3}}$$

In order to oscillate there must be positive feedback from the output to the input (collector circuit to the base). There is a 180 degree phase shift between base and collector. The remaining 180 degree phase shift is provided by the transformer (total 360 degree phase shift). There are a variety of oscillator circuits which use inductor and capacitor tuned circuits.

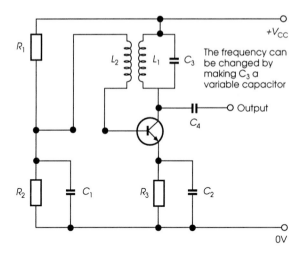

The frequency can be changed by making C_3 a variable capacitor

other in order to maintain oscillation.

An inductive and capacitive circuit is resonant when $X_L = X_c$. This is the circuit's natural frequency of oscillation:

if $2\pi fL = \dfrac{1}{2\pi fC}$ then $f = \dfrac{1}{2\pi \sqrt{LC}}$

A parallel *LC* circuit has an impedance–frequency graph which has a maximum impedance at the resonant frequency. This is a most important fact and forms the basis of not only *LC* oscillators but also the tuning of different channels in radio receivers. At resonance the high impedance of the *LC* circuit provides a load for the transistor which does not exist at other frequencies. Once an electrical pulse starts an oscillation in the *LC* circuit, the transistor needs only to provide sufficient power to overcome the losses in order to maintain the oscillation. In the circuit diagram of Figure 4.13 the necessary 360 degree phase shift, to ensure the feedback signal has the correct phase at the input, is obtained by the 180 degree phase change between the base and collector and the additional 180 degree phase change between the primary and secondary windings of the transformer.

In the circuit:

L_1, C_3 control the frequency of oscillation.

R_1, R_2 control the biasing of the transistor.

C_1, C_2 are decoupling capacitors to prevent signal losses at the oscillator frequency across R_2 and R_3.

C_4 is a coupler to prevent the output circuit affecting the DC conditions of the oscillator.

R_3 is for temperature stability. If the transistor current increases with temperature the voltage rises across R_3. Its polarity is such as to oppose the increase in current by increasing the voltage on the emitter which reduces the base–emitter voltage.

A 180 degree phase shift is provided by the transistor and a further 180 degree phase shift by the three section network. Each section provides a 60 degree phase shift at one particular frequency which becomes the oscillating frequency. Various arrangements of the circuit are used.

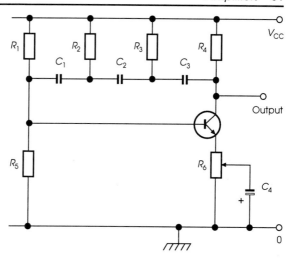

Figure 4.14 *RC oscillator.*

Resistance capacitance oscillators (RC oscillators)

These types of oscillators tend to be used for the generation of low frequencies where inductances and capacitors would be large in LC tuned oscillators. The necessary phase shift is obtained by cascading RC components so that each combination provides a phase shift. In practice each RC combination provides less than a 90 degree phase shift and, therefore, three RC networks, each providing a 60 degree phase shift, are used in order to produce the necessary 180 degree phase shift for the frequency determining unit. This, together with the 180 degree phase change between the base and collector of the transistor, provides the necessary 360 degree phase shift between the input and output for positive feedback. The circuit, therefore, oscillates at the frequency at which the complete RC network provides a 180 degree phase shift. The phase shift depends upon the value of the resistance and the reactance of the capacitor which is frequency dependent. The correct phase shift will, therefore, only occur at one frequency and this becomes the frequency of oscillation.

Wien network

A different approach to the design of RC oscillators requires a non-inverting amplifier, which can be two inverting amplifiers in cascade. As this provides the necessary 360 degree phase shift the RC network must not provide any additional phase shift. Such a network is shown in Figure 4.15. The frequency which makes the impedance of the series RC network equal and opposite to the parallel RC network is the frequency of oscillation:

$$f_r = \frac{1}{2\pi CR}$$

It is a condition for oscillation that the amplifier has a gain of at least three.

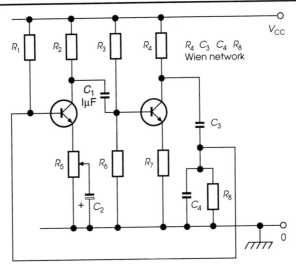

Wien network oscillator

Figure 4.15 *A 360 degree phase shift is provided by the two transistors. The feedback circuit must, therefore, produce no phase shift. The zero phase shift occurs at a particular frequency when the series RC and the parallel RC are all of equal and opposite value.*

Crystal oscillators

For applications requiring the highest stability a crystal oscillator is used. The circuit relies on a piezo-electric effect whereby a quartz crystal can be cut in such a fashion that when a voltage is applied across the faces of the cut crystal it vibrates at a frequency depending upon the cut. Each crystal is cut for a specific frequency and because high frequencies require the crystal to be cut very thinly there is an upper limit of approximately 15 MHz after which it becomes mechanically unstable. In order to generate higher frequencies, harmonics of the fundamental are produced. These can be produced by overtone crystals which operate on an odd harmonic of their fundamental or a crystal is operated at a lower frequency and frequency multiplier circuits are used. For greater stability of frequency the crystal can be enclosed in an oven so that variations do not occur due to temperature changes.

The crystal, when used as an oscillator, has two equivalent circuits, one of which is series resonant and one of which is parallel resonant. The parallel circuit produces the higher resonant frequency. The crystal can be used in either mode but obviously the circuit arrangements are different.

Crystals are cut in a number of ways and each has its own identifying letters. These letters relate to the frequency and temperature stability (typical A–T, B–T; see the section on carrier generation in Chapter 5).

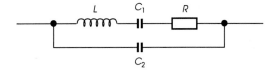

(a) The crystal has two equivalent circuits. There is a series resonance consisting of L, C_1 and R and a parallel resonance which is L, C_1 and R in parallel with C_2.

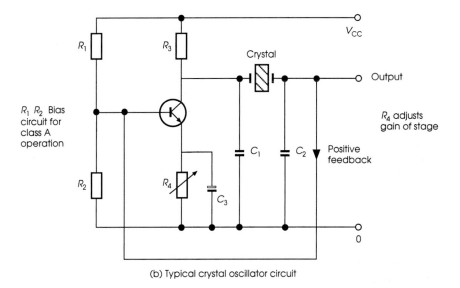

(b) Typical crystal oscillator circuit

The crystal is operating in a series resonant mode. A 180 degree phase shift is provided by the transistor between base and collector and a further 180 degree phase shift is provided by the network of the crystal. C_1, C_2 ensure crystal operates in series mode.

Figure 4.16 *Crystal oscillators.*

5 Transmitters

The theory of radiation, modulation, oscillators and amplifiers given in the previous chapters is combined in the specification and design of transmitters. Transmitters may be required to operate on a single channel frequency which does not change. However, transmitters operating in the high frequency band may have to change frequency with changing conditions in the ionosphere in order to maintain communication. Mobile transmitters may have to be able to switch frequency to a particular channel within a band in order to obtain a free channel. These requirements affect the complexity of the transmitter. In addition to the possible requirement for frequency changing, power output and the operating frequency vary considerably in transmitters used for different purposes. However, the basic building blocks remain the same although circuitry can be very different.

The actual design of transmitters shown in Figure 5.1 is determined by:

1. The frequency of operation.
2. Any requirement to change the operating frequency.
3. The power output required from the transmitter.
4. High or low level modulation techniques.
5. Type of sideband operation.
6. Bandwidth of the transmitted signal. (Communication transmitters do not need the noise and distortion specification of a broadcast transmitter.)

Only amplitude modulation tends to be used at the lower radio frequencies due to the wide bandwidth required to transmit frequency modulated signals. This is not available at low and medium radio frequencies where channels are spaced close together and signals can travel long distances. In the VHF and UHF frequency bands both amplitude and frequency modulation are used as signals do not travel long distances and greater bandwidth is available for each channel.

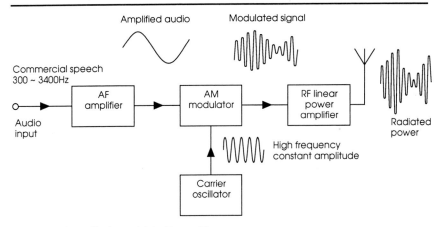

(a) Low level amplitude modulated transmitter

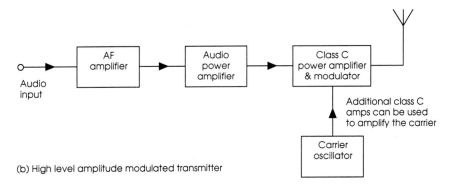

(b) High level amplitude modulated transmitter

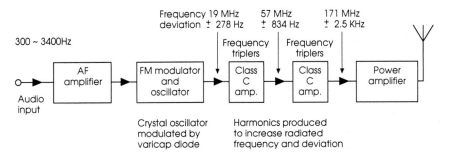

(c) Typical mobile radio FM transmitter

Figure 5.1

Carrier generation

Because of the accuracy required, the carrier frequency is generated with a crystal (Chapter 4) and these can be cut readily to give a typical stability of ±0.005% over a wide temperature range and for frequencies between 2 and 15 MHz it is possible to obtain a precision A–T cut with limits of ±0.0015% over a temperature range of −20 to +70 °C. The carrier is generated at a constant frequency and amplitude. The generated frequency from the crystal is chosen as a sub-multiple of the required frequency and the crystal oscillator is followed by a series of frequency multipliers. The crystal can also be operated in overtones where harmonics of the crystal fundamental are generated. For low frequency transmitters the crystal frequency can be higher than required and this can be divided. This method provides improved stability.

The frequency multiplier stages are class C amplifiers in which the amplifier, because of its method of operation, contains not only the input signal frequency but also harmonics of the frequency. The load of the amplifier is a tuned circuit which, instead of being tuned to the input signal, is tuned to a harmonic and this new frequency becomes the output signal. Small multiplications, which are usually less than four, are made in each stage and, should a greater frequency be required, additional frequency multiplier stages can be cascaded. This is required because of the lower power in the higher harmonics.

Multiple carrier frequencies

When a communication transmitter operates in the high frequency band for international radio telephony there is a need to change frequency, as the ionosphere conditions change, if continuous reception at any particular point is to be maintained. If crystals were used for the individual frequencies this would lead to a large number of crystals and considerable time to change the system manually. Modern communication transmitters are self-tuning once the required operating frequency has been chosen by the operator. The system is operated by mixing the channel modulated frequency from a drive unit (see later) with frequencies derived from a 'frequency synthesiser' to produce the required output frequency. The output stages are then automatically tuned by motor driven variable inductors and capacitors to the new transmitter frequency.

Although overtone crystals can be directly operated up to approximately 150 MHz the system is not as stable as using a low frequency crystal followed by frequency multiplication circuits. When using single sideband and independent sideband systems the carrier frequency must be reinserted in the receiver for demodulation and this becomes more difficult if the original carrier is not stable.

In mobile radio systems channel frequencies have to be switched. This requires the generation of many carrier frequencies which must maintain close tolerances as the channels are very close together. The modern frequency synthesiser performs this function as it is able to produce the required channel frequencies correctly spaced under the control of a single crystal.

Frequency synthesiser

The systems used in the past and present are:

1. Sum and difference type: These use a number of crystals which are selected and mixed together to produce sum and difference frequencies. These systems produce not only the required frequency but also unwanted frequencies which can produce interference to different channels. They also still use a number of crystals, which is uneconomic.

2. Harmonic type: This uses only one crystal but uses multiplication and division circuits to produce a considerable number of individual frequencies from the original frequency. These frequencies are applied to a number of mixers to produce the sum and difference frequencies which can be selected as the output channel frequency. Again such a system also produces many unwanted interfering signals from the mixer circuits.

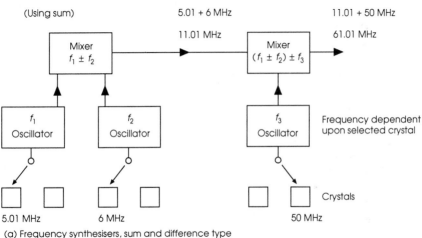

(a) Frequency synthesisers, sum and difference type

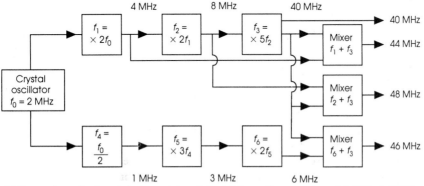

(b) Harmonic synthesiser. A single crystal oscillator is followed by a number of multiply or divide circuits. The outputs of these circuits are applied to mixers to provide the sum of the individual frequencies

Figure 5.2   *Frequency synthesisers.*

3. Phase locked loop synthesiser: This is the more modern system and uses integrated circuits to produce the phase locked loop circuit. The system uses a voltage controlled oscillator whose frequency is dependent upon the value of a capacitance

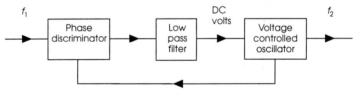

Figure 5.3   *Phase locked loop. The phase discriminator compares frequency and phase of f$_1$ and f$_2$. A voltage is produced which is proportional to difference. The voltage controlled oscillator is an LC oscillator with a varicap across the tuned circuit. The output frequency and phase are controlled by DC voltage. DC voltage will vary, causing change in f$_2$ until the phase of f$_1$ and f$_2$ are the same. There is a capture range over which f$_2$ can lock to f$_1$.*

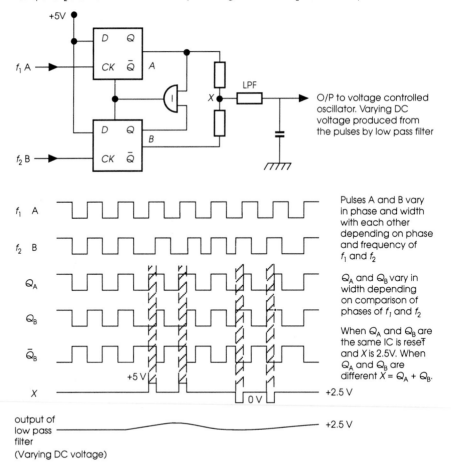

Figure 5.4   *Digital phase discriminator circuit. The frequencies are applied to an integrated circuit in which pulses produced from the frequencies are compared.*

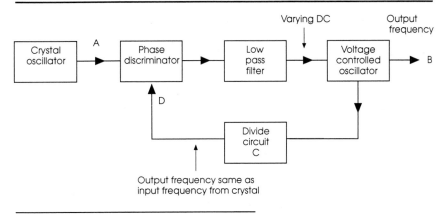

Examples	1	2
A	100 kHz (crystal)	100 kHz
B	100 MHz (output frequency)	1.5 MHz
C	÷ 1000 (divide circuit)	÷ 1050
D	100 kHz (same as crystal)	100 kHz

Figure 5.5 *Changing the frequency. If a variable divide circuit is inserted in the path of the pulses being fed back, it is possible to change the ouput frequency of the voltage controlled oscillator by altering the count.*

across the tuned circuit. The capacitance is a varicap whose value depends upon the DC voltage applied to it. The oscillator output frequency, therefore, depends upon the voltage fed to the varicap.

The crystal frequency runs at a fixed frequency and its output is fed to the phase discriminator together with the feedback circuit output from the voltage controlled oscillator. A variable divide circuit is incorporated in the feedback circuit so that at the phase discriminator both the output from the crystal circuit and the voltage

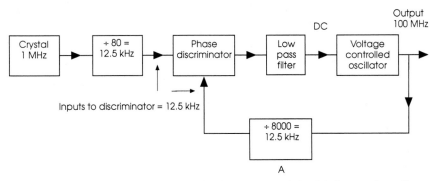

Figure 5.6 *Mobile radio channel spacing. By altering the count circuit A, frequencies on the output will change and provide 12.5 kHz between channels.*

controlled oscillator appear to be the same. When this occurs the voltage applied to the varicap is constant and the frequency of the voltage controlled oscillator remains stable. However, if the count in the divide circuit changes, the frequencies to the phase discriminator will differ and a different voltage will be applied to the varicap. This changes the output frequency until the two frequencies at the phase discriminator are again in phase and the voltage to the varicap becomes constant. The output frequency from the voltage controlled oscillator remains stable and if it tries to drift, a DC voltage change from the phase discriminator is produced which corrects the error. In private mobile radio (PMR) the channels are 12.5 kHz apart and, therefore, the frequency at the phase discriminator is arranged by divide circuits to be 12.5 kHz. Changes in the count circuit then produce different output frequencies from the voltage controlled oscillator for the channels which are 12.5 kHz apart.

The phase discriminator produces pulses from the signals applied to its inputs. The pulses are compared within the IC and differences in phase produce variable pulses in amplitude and frequency depending upon the differences between the signals. When in phase a mid-range voltage is produced on the output of the IC. The output pulses are passed through a low pass filter to produce a varying DC voltage before applying it to the varicap.

High and low level modulation

When the carrier has been generated, it must be both modulated with the information signal and amplified to provide the power required on the output of the transmitter.

Modulation of the carrier can be performed at either the low or the high power stages. These are:

1. The initial stages of the transmitter when the carrier is at a low level and, therefore, the modulating signal need also only be at a low level. This has the advantage of simplifying the modulating signal circuitry and little amplification is required. When the carrier is amplified, in order to obtain the required power output, class B amplification is used in order not to distort the modulated signal. (Class C tuned amplifiers introduce severe distortion when amplifying amplitude modulated signals.)

2. The carrier can be amplified to provide sufficient power for transmission and the modulation process can take place in the latter stages of the transmitter. This has the advantage that class C amplification can be used to amplify the carrier which provides maximum efficiency. However, the modulating signal must also be amplified to provide sufficient power for the modulating process. This invariably must be done with class B amplifiers in order to obtain sufficient power economically.

High level modulation is used mainly with:

1. Double sideband amplitude modulated transmitters used for broadcasting.
2. VHF and UHF mobile transmitters.

Low level modulation is used mainly with:

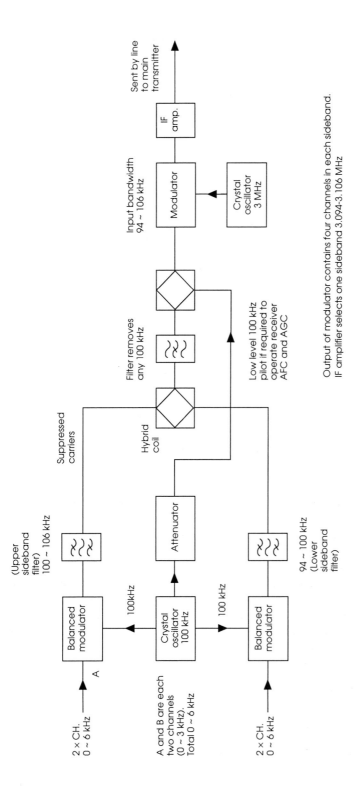

Figure 5.7 *Drive unit for independent-sideband amplitude modulation system (ISBSC).*

1. Single sideband transmitters (SSBs, see Chapter 3).
2. Independent sideband transmitters (ISBs).

In these latter transmitters the modulation process is performed in a drive unit which can be separate from the transmitter. The transmitter, therefore, provides the transmitted power and the transmitted frequency. The ISB systems are used for international radio telephony links in the high frequency band. Drive units for single sideband suppressed carrier systems are used in maritime mobile systems.

Drive unit

As described in Chapter 3, ISB working involves a common carrier frequency but the upper and lower sidebands are produced by different signal inputs (*A* and *B*). In the drive unit the two signals *A* and *B* are applied to separate balanced modulators fed with the same crystal derived carrier frequency of 100 kHz. Each of the signals *A* and *B* has a 6 kHz bandwidth and has been previously multiplexed to contain two channels of commercial speech each of 3 kHz bandwidth. The outputs, therefore, have suppressed carriers and upper and lower sidebands of 94 to 100 kHz and 100 to 106 kHz. Both *A* and *B* modulated signals are passed through band-pass filters which remove either the upper or the lower sideband. The two remaining sidebands are combined and, if required, a low level pilot carrier can be reinserted to provide a signal for the operation of the automatic gain and frequency control at the receiver.

The two sidebands are now modulated onto a 3 MHz carrier derived from a crystal oscillator. The sideband frequencies are, therefore, translated to 3.094 to 3.106 MHz. It is these frequencies which are sent by line to the transmitter for power amplification and further frequency translation for radiation in the high frequency band of 4 to 30 MHz.

Drive units for single sideband suppressed carrier systems are simpler but similar. There is no combining unit and the input modulating signal is a single audio signal of 3 kHz bandwidth. This is applied to the balanced modulator which is supplied with 100 kHz to produce the sidebands of 97 to 100 kHz and 100 to 103 kHz. Only the lower sideband is passed by a bandpass filter. The lower sideband is again mixed with a 3 MHz frequency to translate it to 3.097 to 3.1 MHz which is sent to the transmitter for power amplification and frequency changing to the allocated radiated frequency in the high frequency band.

Amplitude modulators

The modulator circuits used for double sideband amplitude modulated transmitters are usually either anode or collector modulated class C tuned amplifiers. However, any non-linear device such as a diode or a transistor working on the non-linear part of the curve relating input voltage and output current will produce sum and difference frequencies if two independent frequencies are applied in series.

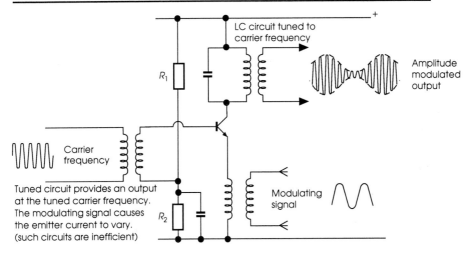

Figure 5.8 *Amplitude modulator.*

It is important that only the wanted frequencies are produced on the output and not additional harmonics as these may cause interference problems. In addition different circuits known as balanced modulators must be used if the carrier has to be suppressed.

Double sideband modulators

A non-linear transistor type modulator is given in Figure 5.8. The bias is adjusted to provide the necessary non-linearity of the input voltage and collector current relationship. The load is a tuned circuit designed to be resonant at the carrier frequency. Without any modulating signal input a sine wave output waveform is produced by the tuned circuit and carrier input signal.

When a modulating signal is applied to the emitter, the current in the collector varies and the carrier and sidebands (f_c+f_m f_c f_c-f_m) appear in the load. Other unwanted frequencies are also produced, but, because the load is a tuned circuit, whose bandwidth can be adjusted to accept only the required frequencies, the unwanted frequencies are rejected. These types of modulators create distortion, are inefficient and can only be used for low power applications.

For high level modulation, anode modulated class C amplifiers are used. To drive such stages the modulating signal is invariably amplified by a class B push–pull amplifier. The output of this amplifier is coupled into the anode (valves), or collector when transistors are used in small transmitters, by a transformer. Any variation in anode/collector voltage, caused by the modulating signal, causes the carrier signal in the tuned circuit to vary in sympathy with it. Again the tuned circuit has a band-width which rejects any unwanted frequencies and only the carrier and sidebands appear across the load.

Sometimes a single stage of modulation is insufficient to obtain the required depth of modulation and in these instances the modulating signal output amplifier drives the last two class C amplifiers by using a common inductor to the collectors of the two stages. This inductor is the secondary of the transformer of the class B amplifier across which is the modulating signal which varies the voltage on the collectors (Figure 5.10).

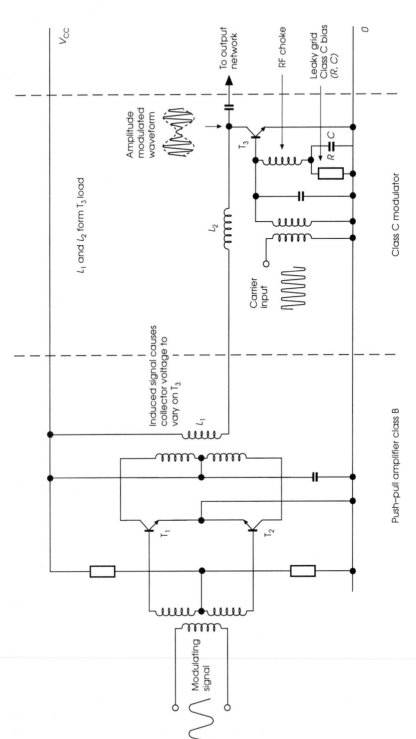

Figure 5.9 *Collector modulated class C amplifier.*

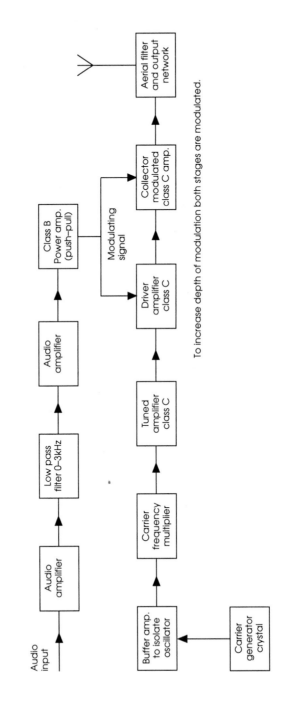

Figure 5.10 *Amplitude modulated VHF mobile transmitter (see also page 78).*

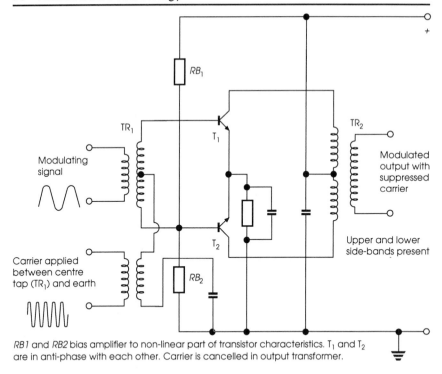

RB1 and *RB2* bias amplifier to non-linear part of transistor characteristics. T_1 and T_2 are in anti-phase with each other. Carrier is cancelled in output transformer.

Figure 5.11   *Transistorised balanced modulator.*

Amplitude modulation suppressed carrier

A similar circuit to the push–pull amplifier can be used with the modulating signal applied in anti-phase to the base of each transistor via the input transformer. Simultaneously the carrier is applied in phase to the base of each transistor from the centre tap of the input transformer. Both transistors are biased to the non-linear part of their characteristics and, therefore, the sum and differences frequencies appear in the output. However, the carrier frequency is cancelled in the output. In a practical circuit, because of different characteristics in the two halves of the circuit, the carrier will not completely cancel and some will appear on the output.

Switching diode modulators

A different approach is to use the diode as a switch. Figure 5.12 shows the single balanced diode modulator. The carrier, which must be much larger than the modulating signal, is applied to the centre points of the diode circuit. The modulating signal is applied to the input transformer. The carrier voltage causes the diodes to switch on and off and the modulating signal appears in the output as a chopped signal. However, it is found that the output signal contains the upper and lower sidebands and the modulating signal but not the carrier. Again due to imbalance and the diode characteristics, carrier leak occurs and unwanted side frequencies are produced.

Carrier switches diodes into conducting and non conducting states. In diode modulators the carrier voltage must be much higher than modulating voltage.

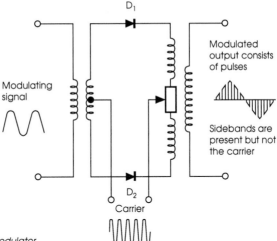

Modulated output consists of pulses

Sidebands are present but not the carrier

Figure 5.12 *Single balanced modulator.*

This is a less expensive system as it does not use transformers. When the diodes are conducting they short circuit the modulating signal. When not conducting the waveform passes to the output. The modulating waveform is therefore chopped.

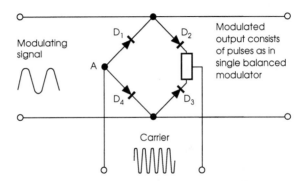

Modulated output consists of pulses as in single balanced modulator

Carrier switches diodes into conducting and non conducting states.

Figure 5.13 *Cowan modulator.*

A less expensive circuit which does not need transformers is the Cowan modulator. This consists of four diodes arranged in such a way that the applied carrier wave switches them into a conducting state on one half cycle and off on the other half cycle when they become reversed biased.

The diode circuit is across the input to output path of the modulating signal. The modulating signal is, therefore short circuited or passed through to the output depending upon the bias on the diodes. Again the modulating signal is chopped and an analysis of the output waveform shows it has the required upper and lower sidebands and modulating signal but no carrier signal.

Balanced modulators are available as integrated circuits. These circuits offer simplicity in use and good electrical stability. They provide fully balanced input and output circuits and a very good carrier suppression over a range of frequencies up to approximately 100 MHz.

Mobile VHF transmitter (amplitude modulated)

Figure 5.10 shows the arrangement of a typical VHF transmitter. If the transmitter has to change frequency then the carrier is generated by a crystal and the required frequencies are produced by frequency synthesis. Single frequency transmitters will use either overtone crystals or a low frequency crystal followed by frequency multipliers as previously explained. Most VHF mobile transmitters have a relatively low wattage output and are, therefore, normally designed with solid state electronics. High level modulation is normal and class C tuned amplifiers are used to produce the carrier power. In order to provide sufficient depth of modulation the output and penultimate amplifiers can be simultaneously modulated as previously described.

The microphone speech channel is band limited to 3 kHz and, because of high level modulation, a class B amplifier is used to drive the collector modulated class C amplifier. An aerial filter is used to limit the frequency output of the transmitter in order to avoid interference with adjacent or other channels caused by the generation of spurious frequencies.

Frequency modulated transmitters

Frequency modulation has the advantages of less interference from amplitude noise and it is radiated as a constant amplitude carrier. This allows the transmitters to provide constant power. However, it has the disadvantage of requiring a greater bandwidth than amplitude modulation in order to carry the same information. Frequency modulation is used for television sound broadcasting, VHF sound broadcasting, SHF radio relay links and VHF and UHF mobile radio systems. It is used at frequencies where radio waves which are transmitted from an omni-directional aerial do not travel long distances and there is more bandwidth available for each channel.

Figure 5.14 shows a typical frequency modulated transmitter system in which a conventional *LC* oscillator is altered in frequency by modulating voltages which are applied to a varactor diode. The FM signal is amplitude limited and then the

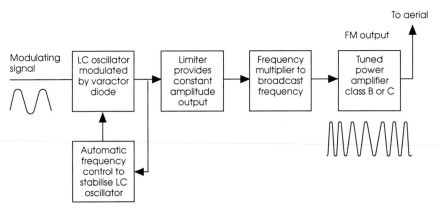

Figure 5.14   *Typical sound FM broadcast VHF transmitter.*

frequency is multiplied to the required output channel frequency. A power amplifier of either class B or class C is used to provide the radiated power. In order to improve the stability of the *LC* oscillator automatic frequency control is included to stabilise the unmodulated carrier frequency.

An improvement in oscillator frequency stability is achieved by using a crystal as the carrier source but frequency modulation of the crystal only succeeds in producing a very small frequency change of 100 Hz or less. In order to improve the deviation of the system, the modulation process is followed by several stages of frequency multiplication. Frequency multiplication not only increases the carrier frequency but increases the deviation by the same multiplication ratio. Such systems are used in narrow band mobile transmitters (Figure 5.1(c)).

6 Receivers

When Marconi and then the BBC started broadcasting, the majority of people listened to them on crystal sets until valve sets became more readily available. All that is required to receive an amplitude modulated radio is a tuned circuit whose resonant frequency is the carrier wave, a diode to demodulate the AM signal and headphones. A long wire acts as an aerial and the set is earthed, possibly to the cold water tap.

With the introduction of amplification the simple radio receiver consists of an RF amplifier with a tuned circuit as the load, a diode demodulator, an audio frequency amplifier and a AF power amplifier to drive the loudspeaker. Different stations can be selected by altering the value of the variable capacitor and tuning the resonant frequency to different carrier frequencies. These types of radio sets are known as TRF sets (Tuned Radio Frequency). Although they receive the different stations such simple receivers would not be used today. The problems are:

1. When the receiver is working in an area where many stations are transmitting, the selectivity of the simple tuned circuit is insufficient to stop adjacent stations being heard simultaneously with the required programme. Additional tuned circuits

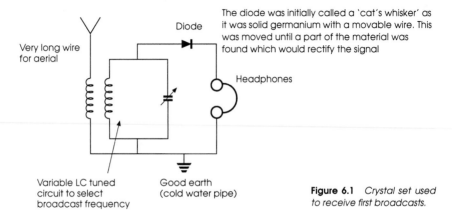

Very long wire for aerial

Diode

The diode was initially called a 'cat's whisker' as it was solid germanium with a movable wire. This was moved until a part of the material was found which would rectify the signal

Headphones

Variable LC tuned circuit to select broadcast frequency

Good earth (cold water pipe)

Figure 6.1 *Crystal set used to receive first broadcasts.*

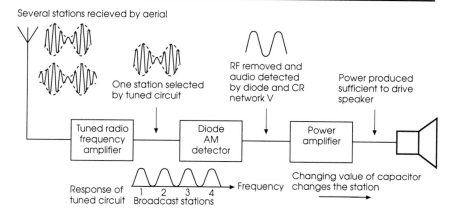

Several stations recieved by aerial

One station selected by tuned circuit

RF removed and audio detected by diode and CR network V

Power produced sufficient to drive speaker

Tuned radio frequency amplifier		Diode AM detector		Power amplifier

Response of tuned circuit
1 2 3 4
Broadcast stations
→ Frequency

Changing value of capacitor changes the station

The tuned circuit (LC) is the load on the amplifier and only the frequency response of the tuned circuit is amplified. By altering the value of the capacitor the resonant frequency can be changed and a different broadcast station selected. (See also Figure 4.12, page 59.)

Figure 6.2 *Simple tuned radio frequency receiver (TRF). (amplitude modulated signals)*

can be incorporated but this requires the mechanical ganging together of the variable capacitors which can then be operated simultaneously. Each capacitor must have the same value in order to tune the circuits to the same frequency and this becomes difficult due to stray capacity. Each tuned circuit also requires electrical isolation from the others and this involves a separate amplifier for each circuit with the possibility of introducing feedback and oscillations.

2. The gain of the radio set depends upon the frequency of the received station. This is because the tuned circuit is the load of the amplifier and its impedance value changes with the changing value of either the inductance or the capacitance when retuning the receiver. The impedance of the parallel tuned circuit at resonance is L/CR ohms. If either the inductance or the capacitance varies then the impedance varies and the gain of the stage alters. Altering the gain also affects the shape of the resonance curve and the selectivity of the stage alters. The resonant frequency of the LC circuit is normally changed by the capacitance but the inductance can be used. It can be seen by the ratio L/C (R is negligible) that as the value of the capacitance increases the gain of the stage decreases, but if the inductance is increased the gain increases. Both increases in value have the same effect of reducing the resonant frequency (see page 60).

Superheterodyne receiver

To overcome the problems of variable gain and poor selectivity the superheterodyne receiver was developed and this forms the basis of modern receiver design. The receiver uses the principle of heterodyning to change all incoming radio signals to the same frequency. This is known as the intermediate frequency (IF) and this stage is designed for a high gain and good selectivity. Because the intermediate

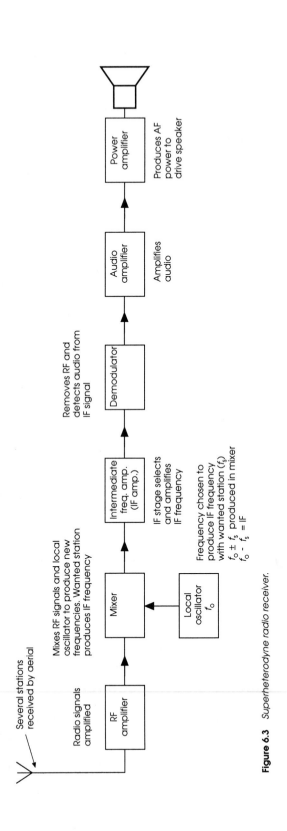

Figure 6.3 *Superheterodyne radio receiver.*

frequency stage provides most of the gain of the receiver, selectivity and gain are not dependent upon the frequency of the received signal. As the IF frequency is constant and is not altered when tuning to different broadcasting stations, the stage can be designed for optimum gain and bandwidth which is then the same for all stations. In communication receivers it may be an advantage to alter the bandwidth of the stage depending upon the type of signal being received.

Unfortunately the design of a superhet receiver introduces problems which are not present in TRF receivers as they are solely the result of producing an IF frequency. The choice of the intermediate frequency is important when considering these problems but it usually has to be a compromise in order that all the operating parameters are taken into consideration. These will be explained as each problem is considered.

Producing an intermediate frequency

In Chapter 3 it was shown that when two signals with different frequencies are applied to a mixer, new frequencies are produced which are the sum and difference of the two frequencies. If f_1 and f_2 are applied, the output of the mixer will contain f_1+f_2 and f_1-f_2. This principle provides the means to change the frequency of any incoming signal to the required intermediate frequency by heterodyning it with a locally produced frequency within the receiver. Each incoming signal is tuned for reception by altering the frequency of the local oscillator until the intermediate frequency is produced. A locally produced frequency above or below the incoming signal is capable of providing the IF frequency by selecting either the sum or difference frequency. The frequency of the local oscillator can be either f_{sig} + IF or f_{sig} – IF. In practice the local oscillator has a higher frequency than the incoming signal and the difference frequency is selected. The reason for this is that the local oscillator is tuned by a variable capacitor which must change its value in order for the local oscillator to produce the intermediate frequency throughout the band of frequencies being received. The ratios of the capacitance values to provide the lowest and highest local oscillator frequencies are different for the two possibilities of using either $f_{osc} = f_{sig}$ + IF or $f_{osc} = f_{sig}$ – IF. For example, assume an IF of 500 kHz and a band of incoming signals of 600 kHz to 1000 kHz. The local oscillator could be either:

$$f_{osc} = f_{sig} + IF \quad \begin{aligned} f_{osc} &= 600 \text{ kHz} + 500 \text{ kHz} = 1100 \text{ kHz} \\ f_{osc} &= 1000 \text{ kHz} + 500 \text{ kHz} = 1500 \text{ kHz} \end{aligned}$$

to

or:

$$f_{osc} = f_{sig} - IF \quad \begin{aligned} f_{osc} &= 600 \text{ kHz} - 500 \text{ kHz} = 100 \text{ kHz} \\ f_{osc} &= 1000 \text{ kHz} - 500 \text{ kHz} = 500 \text{ kHz} \end{aligned}$$

to

It can be seen in this example that the tuning range required of the local oscillator using $f_{osc} = f_{sig}$ + IF is 1500/1100 which is approximately 1.36.

When using a local oscillator whose frequency is $f_{osc} = f_{sig}$ – IF the tuning range is 500/100 which is a ratio of 5. Although there is no significance to the figures used as they were chosen for simplicity, the same principle applies for any situation. The higher ratio makes the tuning of the local oscillator more difficult and expensive.

Image frequency

In the previous section it has been shown that the local oscillator frequency is placed above the incoming signal by a frequency difference equal to the IF frequency. However, should a signal with a frequency equal to f_{osc} + IF appear at the mixer then an IF frequency is produced and the unwanted signal will interfere with the required channel. This unwanted signal is known as an image frequency and always appears at a frequency which is 2 × IF from the required channel. The mixer stage cannot reject it and, therefore, it must be stopped in the stages prior to the mixer. This is the radio frequency amplification stage and the tuned circuit in this stage must be able to reject the image frequency. This is easier if the image frequency is far removed from the required channel and for this purpose the IF stage should be a high frequency. The ratio of the IF frequency to the signal frequency is greater at the lower signal frequencies and is reduced as the signal frequencies increase.

This makes the rejection of the image frequency more difficult at the higher frequencies and, therefore, different IF frequencies are chosen for receivers working in different bands. Receivers working in the higher bands will have a higher intermediate frequency. However, as will be seen, this has an adverse effect on adjacent channel interference. If any of the image frequency passes to the mixer it will cause crosstalk and if a signal, which is a few kHz from the image frequency, reaches the mixer it will result in two frequencies being present in the IF which beat together and cause whistles in the loudspeaker. The interference from the image is given as the image response ratio and is quoted in decibels. It provides a ratio of the voltages of the required signal and the image signal at the receiver input to provide the same output.

Adjacent channel

The ability of the receiver to reject the adjacent channel is mainly determined by the frequency response and gain of the intermediate frequency amplifier and because this amplifier remains constant its characteristics can be designed for optimum performance. Different types of receivers require a different bandwidth response and these are defined as the −3 dB points on the output voltage/frequency curves. It is the shape of these curves which is important and often the frequencies relating to the −6 dB and −60 dB points are given so that the shape can be established. These points are often known as the nose and skirt bandwidths respectively and their frequency ratio is known as the shape factor. Frequency modulated signals used for broadcasting require approximately 180 kHz bandwidth while AM broadcasting (double sideband) requires approximately 9 kHz. Communication receivers vary depending upon the type of sideband operation. A single sideband receiver requires approximately 3 kHz bandwidth.

The specification for adjacent channel interference is given by the adjacent channel ratio. This is the ratio in dB of the input voltages at the required frequency and that at the adjacent channel which produces an output power for the adjacent channel which is -30 dB that of the required signal power:

$$\text{Adjacent channel ratio} = 20 \log\left(\frac{\text{adjacent ch. voltage}}{\text{required ch. voltage}}\right)$$

where the adjacent channel voltage produces −30 dB of the power of that of the required channel voltage.

To provide adequate adjacent channel selectivity a low intermediate frequency is desirable but this conflicts with the requirements for image channel rejection. The IF amplifier frequency must, therefore, be a compromise. There are, however, other factors to consider. It is desirable that the IF is outside the tuning range of the receiver. This allows an IF trap to be used in the RF stage to prevent any signal at the intermediate frequency which is picked up by the aerial from reaching the mixer stage and causing interference. A parallel tuned circuit tuned to the intermediate frequency can be inserted in series with the aerial or a series circuit to earth can be used to short-circuit the signal.

The lower the intermediate frequency the easier it is to design the intermediate amplifier for gain. However, the bandwidth of the coupled–tuned circuit used in such amplifiers is proportional to its resonant frequency and, therefore, if a large bandwidth is required, as in FM transmission, the higher must be the intermediate frequency. Typical values are 465 kHz for broadcast receivers using AM modulation and 10.7 MHz for FM modulation.

Radiation

The local oscillator which is used for the mixing process is itself an RF signal which can cause interference to other equipment if it is allowed to radiate from the receiver. It must be prevented from coupling into the aerial by the use of an RF amplifier and physical screening of the oscillator may be necessary.

Additional interfering signals

Although adjacent channel and image channel interference have been specifically detailed, the superhet receiver is also susceptible to other interference problems. If two or more signals appear at the aerial which are separated by or close to the intermediate frequency, any non-linearity in the RF or mixer stage produces intermodulation products whose frequencies pass through the IF stage as interfering signals. It is essential that the gains of these stages are such that non-linearity is avoided.

Co-channel interference is a further problem if, due to unusual propagation conditions, a channel operating on the same frequency is picked up by the aerial. Under these conditions there is nothing one can do unless directional aerials can be used and the transmissions are from different directions.

A further problem can be caused by harmonics of the oscillator combining with stations whose frequency is such as to produce intermodulation frequencies which fall within the passband of the IF amplifier and cause interference signals.

In an amplitude modulated system, if non-linearity occurs in the RF or mixer amplifiers, cross-modulation can occur whereby the AM modulation from an unwanted signal can be transferred to the required signal when they are both present at the aerial as relatively strong signals. This is prevented by ensuring that the selectivity is sufficient to prevent large unwanted signals entering the receiver stages and also ensuring that the RF stage is operating under linear conditions.

Intermediate frequency amplifiers

The construction of these amplifiers traditionally used transformers as the load of the amplifier. The primary and secondary couple the stages of the amplifier and each is tuned with a capacitor to the intermediate frequency. The coupling between the primary and secondary is adjusted until critical coupling is obtained and this produces a gain/frequency response for the stage which has a flat top with steep sides. This type of response allows linear amplification throughout the required bandwidth and good selectivity.

Modern amplifier designs use high gain amplifiers with a wide bandwidth. The shaping is controlled by a ceramic or crystal filter. This has the advantage that no setting up is required as there are no adjustable parts. Fully integrated circuits are also being manufactured which contain the IF amplifier, demodulator, squelch circuit and audio pre-amplifier which have been specially designed for low power narrow band FM communications.

Local oscillators

The local oscillator which feeds the mixer must be both accurate and stable. Tuned LC circuits are used for broadcast receivers in the MW, LW, SW and FM. Crystals and synthesisers are used in communication receivers. In land mobile receivers the local oscillator actually selects the frequency of the received signal. The RF amplifier is tuned to receive all the signals within a particular band but the local oscillator frequency determines which signal produces an intermediate frequency at the mixer. All the other signals are rejected by the IF stage. This provides a system which allows a channel to be selected by selecting a local oscillator frequency.

Double superhet receiver

It has been previously shown that a low frequency IF is desirable to provide rejection of the adjacent channel but a high frequency IF is required for the rejection of the image channel. In the low and medium frequency bands a compromise can be made which is acceptable but in the high frequency bands the problem becomes more difficult. A receiver containing two IF frequencies could, however, satisfy both requirements. A typical mobile radio double superhet is shown in Figure 6.4. The first IF is a high frequency stage in order to provide good separation of the wanted signal from the image signal which allows its rejection in the RF stage. The second IF is a low frequency stage which allows adjacent channel rejection in the tuned IF amplifier.

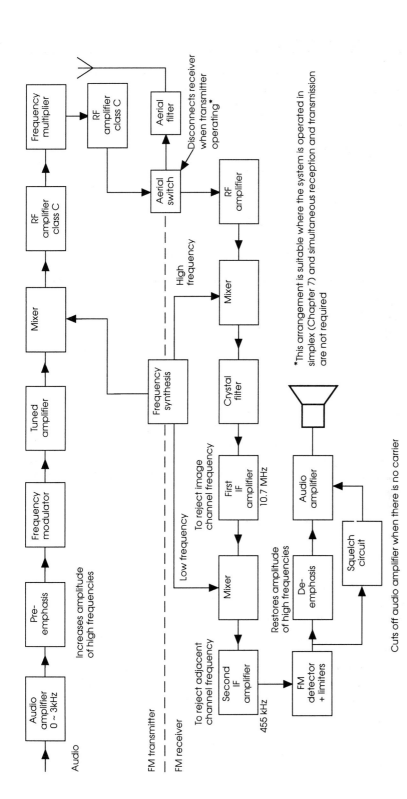

Figure 6.4 *Typical VHF frequency modulated transceiver.*

Demodulation and automatic gain control

Amplitude demodulation

The process of demodulation involves the recovery of the original signal from the RF carrier. The simplest demodulation process is for amplitude modulated double sideband signals. Demodulation is achieved by half wave rectification of the carrier by the use of a diode followed by a resistor and capacitor whose time constant is correctly chosen to recover the audio waveform (see Figure 6.5).

The output of the stage also contains a DC voltage which varies with the amplitude of the signal. The DC is, therefore, related to the received signal and can be used to stabilise signal levels which are received at the aerial. These vary consider-

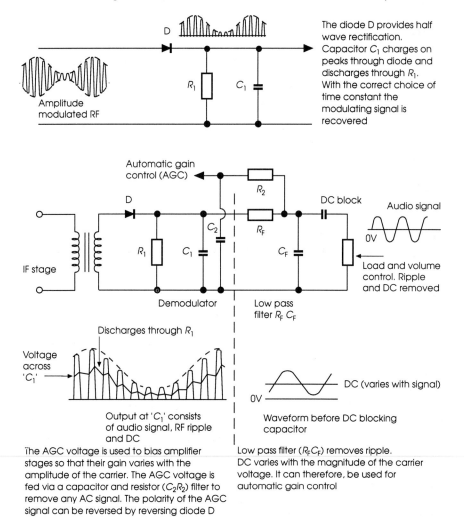

Figure 6.5   *Amplitude modulation detection.*

ably with propagation conditions. Stabilisation is achieved by using the DC voltage as an automatic gain control (AGC) which alters the bias of the previous amplifiers, and therefore their gain, depending upon the strength of the signal. The audio output from the speaker will not depend upon the strength of the signal but only on the modulation depth of the signal.

The use of AGC allows the RF amplifier to be a high gain amplifier for weak signals which is reduced when a large signal is received. This prevents overloading the stage and the production of distortion, harmonics and cross-modulation.

AGC is used in all superhets including FM receivers. In FM receivers the AGC is used to ensure the carrier is sufficient to operate the limiters which remove any amplitude variations and to ensure the RF and IF amplifiers are not overloaded.

Frequency demodulation

In frequency modulation the information is carried in the variations of the carrier frequency. These frequency variations have to be detected and changed into variations of voltage which represent the original amplitude variations of the speech.

The FM demodulators are known as discriminators and there are several designs. Two circuits have been popular with manufacturers and these are the ratio discriminator and the Foster–Seeley discriminator. In recent times integrated circuits have become available which incorporate the FM discriminator and these are used in mobile FM receivers.

Ratio discriminator

Figures 6.6 and 6.7 shows the voltage output from a discriminator circuit with changes in input signal frequency. It is designed that zero voltage occurs when the carrier frequency is the input signal and a positive or negative voltage occurs when the input signal has changed in frequency due to deviation. The voltage changes represent the modulating signal.

For this to occur in a ratio discriminator a transformer is tuned in its primary and secondary circuit to the carrier frequency and this produces a 90 degree phase shift between the input and output signals. This phase will vary if the frequency changes from the resonant frequency.

If the primary voltage is applied in addition to the centre tap of the secondary winding, voltages at either end of the secondary winding are the vector addition of the primary and secondary voltages. As the induced secondary voltage is varying in phase with the changing frequency, the voltages at the top and bottom of the secondary winding change in amplitude with frequency changes. Two amplitude demodulators are used to detect the changes and an output, which is the combined vector voltages, is obtained which follows the frequency deviation of the carrier. This is the modulating signal.

The output resistors are shunted by an electrolytic capacitor across which is a constant voltage. The time constant of the capacitor and the resistors is important as it must be sufficiently long to prevent any variations caused by amplitude changes in the modulating signal. These would normally be caused by noise and, due to the time constant, would be ignored.

The ratio discriminator is the most commonly used FM detector in AM/FM domestic receivers as it provides self-limiting, which means it removes all amplitude variations to the waveform caused by noise. A Foster–Seeley circuit, although it produces less distortion, is not self-limiting and requires an extra stage to perform this function. The operation of this type of circuit again uses phase differences in a similar manner as the ratio discriminator.

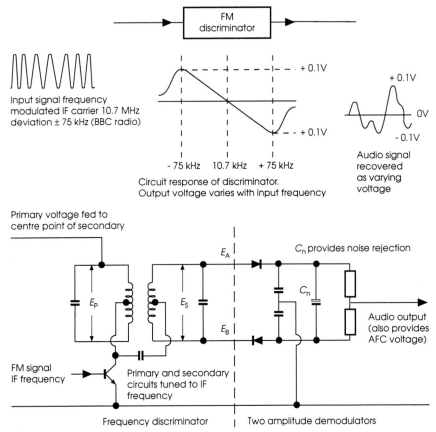

Figure 6.6 *Ratio discriminator.*

Automatic frequency control

It can be seen in Figure 6.3 that the IF stage with its local oscillator precedes the detector stage. If the local oscillator changes frequency, due to any instability, the intermediate frequency will not be exact and will differ to the tuned stages of the detector. Instead of producing zero voltage at the output of the detector when carrier frequency is received, it produces a DC voltage and, therefore, distortion. This voltage, however, can be used as an automatic frequency control by feeding it to a varicap diode across the local oscillator. This changes the local oscillator frequency

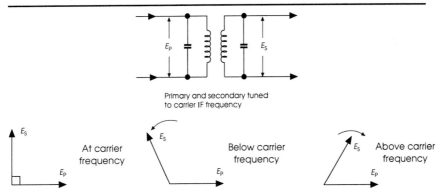

Primary and secondary tuned
to carrier IF frequency

Without centre point connection the phase relationship between primary and secondary
voltages is as above

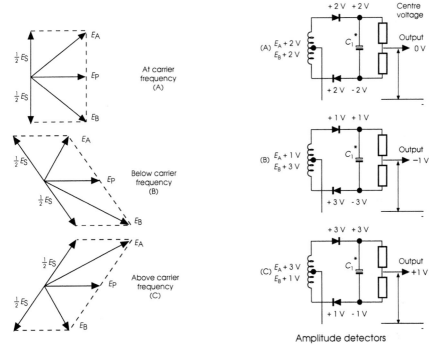

Amplitude detectors

*Constant 4 V across capacitor

With the centre tap the secondary voltages are the vector sum of 1/2 E_S and E_P. These are E_A
and E_B and are applied to the diodes of the amplitude detectors.
   The output of the detector is used for automatic frequency control (AFC) (See text)

Figure 6.7   *Ratio discriminator operations.*

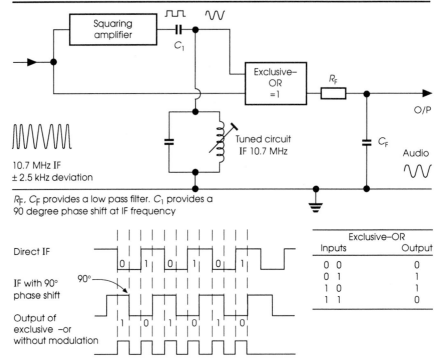

R_F, C_F provides a low pass filter. C_1 provides a
90 degree phase shift at IF frequency

Exclusive-OR	
Inputs	Output
0 0	0
0 1	1
1 0	1
1 1	0

With modulation the phasing between the direct and indirect gate inputs vary. This causes
the pulses from the gate to vary in width. Pulses occur whenever 1 and 0 coincide.

The low pass filter converts the changing DC from the varying widths of the pulses into an
anologue signal

Figure 6.8 *Quadrature discriminator.*

until the correct intermediate frequency is obtained and the detector produces 0
volts at the carrier frequency.

In mobile radio systems crystal oscillators are used which provide high stability
for the local oscillator and, therefore, AFC is not used.

Mobile systems

Integrated circuits incorporating the FM discriminator are commonly used and a
quadrature discriminator is a popular circuit for such purposes. It has the advantages
of using only one tuned circuit, digital techniques and is easy to set up.

A typical block diagram is shown in Figure 6.8 in which the incoming intermediate frequency is fed both to a squaring amplifier and an exclusive-OR gate.
Pulses from the amplifier pass through a capacitor, which introduces a 90 degree
phase shift at the intermediate frequency, to the tuned circuit which is resonant at
the intermediate frequency. This has the effect of making the pulses again sine
waves but differing in phase with the original signal. The phase difference varies

with the signal frequency. The exclusive-OR gate produces output pulses whose widths are dependent upon the phase of the two input sine waves. The DC content, therefore, varies at the audio frequency of the modulating signal. This signal can be recovered by passing it through a low pass filter which removes the IF frequency.

The phase locked loop is also used in integrated circuits and is shown in Figure 5.3. In this circuit the phase of the frequency modulated signal is compared with the frequency of a local oscillator. If the two frequencies vary in phase an error signal is produced on the output which is fed back to the local oscillator. This will change the frequency of the local oscillator until the phases of the two signals are the same. The error signal is, therefore, following the variations in the frequency changes of the incoming signal and can be used as the FM demodulated signal.

Demodulation of single sideband and independent sideband

In order to demodulate these systems it is essential that the carrier frequency is reinserted at the receiver and its frequency must be very accurate. Balanced modulators are used for the demodulation process and the signal and carrier frequency are applied together. As in the modulation process, the reinserted carrier frequency must be at a high level in order to switch the diodes. The sum and difference frequencies are produced and if the received signal is $f_c - f_m$ the lower sideband produced is $f_c - (f_c - f_m) = f_m$. This modulating frequency can be separated from the other sideband products by a low pass filter.

For integrated circuits a heterodyne detector is often used and this can be incorporated with other circuits to form a complete IF amplifier and detection IC. In these circuits (also known as product detector) an active device such as a transistor or FET has the input signal and a local carrier frequency applied to the input. The output current contains the products of the two input signals and a voltage is developed in the load which contains the original modulating signal. This is obtained by filtering it from the unwanted products.

Output audio circuits

In mobile systems the audio stages need only amplify low bandwidth signals of approximately 3 kHz and provide sufficient power for the loudspeaker. These systems usually consist of an integrated circuit which contains both the pre-amp and the power amplifier.

A radio set designed for high gain does not produce an internal AGC when it is not receiving a carrier. Its gain in these circumstances will rise to maximum and the noise which is present will be amplified. The resulting output is very objectionable and circuits are, therefore, included in the design to reduce or cut off the gain when no carrier is received. This is known as the squelch or muting circuit. A typical circuit monitors the carrier at the input to the detector and uses its presence to switch a transistor either on or off and interrupt the audio signal which is fed to the audio amplifier. In some cases the squelch control is made adjustable so that noise is present if desired.

Use is also made of the squelch circuit to operate selected receivers in some types of mobile systems. Signals from a base station can be coded to remove the muting from selected vehicles while others stay muted even though they all receive the signal (Chapter 7).

7 Mobile communication systems

Before the 1980s the mobile radio communication industry in the United Kingdom was limited to the armed services, commercial and public organisations using private systems, and marine and aircraft communication. The general public's first introduction to mobile telephones was the portable telephone. This has its own local base station and the freedom to move around the home or office. Many illegal phones appeared which were not approved for connection to the BT public telephone network. In the 1980s a greater variety of more sophisticated equipment became available and public mobile services were beginning to be used abroad.

Cellular radio

In 1984 mobile phone licences were issued by the government to two companies, Cellnet and Vodafone, for the building and operation of the infrastructure for a public radio telephone network. In the hope of avoiding monopolies separate service providers were established to sell the service, bill and collect the money. The 45 service providers in turn sell the product through dealers who actually deal with the customers. The competition is, therefore, on the method of charging rather than technical facilities.

Cellular radio requires service areas to be arranged into cells which have their own transmitter and receiver base stations. In each cell a group of frequencies are used for communication with mobile car phones and handportables. If a mobile passes from one cell to another there must be a handover of control and a change of frequency which should be undetected by the user. The system is covered in Chapter 10. By 1992 a considerable area of the UK was covered by the systems. In the main population areas reception is good for both car phones and the less powerful handportables. In less populated areas good reception is received by car phones but there is fluctuating reception on handportables. A third area of very low population either has fluctuating reception on car phones and reception is not guaranteed on handportables, or reception is not guaranteed on either type of phone.

Paging systems

Alongside the radio phone are radio paging systems which are separate networks offering facilities which range from making the customer aware that someone is wishing to make contact by receivers which bleep, flash or vibrate to more sophisticated systems which provide visual displays of messages. Some of these systems are providing specialist information such as the latest currency and market reports.

There are two main types of systems. These are on site paging systems, where paging is restricted to a factory, office, hospital, etc., and wide area paging which includes regional and national paging systems. These systems are described in Chapter 9.

Data and fax systems

A facility provided for the users of the cellular system is to have fax messages left for them on a central computer. The customer can phone and decide if the messages are urgent and, if desired, have them printed at a convenient fax machine or phone point to which a portable fax machine can be plugged in. Data, as explained later in this chapter, is a series of digital pulses which analogue circuits cannot handle without a modification to the waveform. The analogue cellular systems and private systems, therefore, need a modem (modulator/demodulator) to convert the digital signals into analogue signals for transmission on the network. The receiving equipment can then be a laptop computer or printer.

Computer data is transmitted on many private radio networks. Many companies such as taxi firms are being actively encouraged to use data and visible messages as the best method of receiving instructions from their dispatcher. This avoids misunderstanding when traffic noise is high or driving conditions difficult. Messages can also be repeated until received correctly.

Public dedicated digital data radio networks have also been provided by several companies which connect a customer's mobile terminal to their own company central computer while travelling throughout the country. A small interface, which provides coding/decoding, a modem and a transmitter/receiver, allows any computer to act as a terminal. This allows computer records to be quickly accessed and immediately updated. Speech is not transmitted on these networks (see Chapter 12). The economics of dedicated digital systems may be difficult as all existing and proposed future systems also transmit data.

Private mobile radio (PMR)

Private mobile radio systems are extensively operated in the VHF and UHF bands by such organisations as the police, ambulance, fire, electricity, water, gas, AA, RAC, taxi firms and many commercial organisations.

As monochrome television is no longer transmitted the frequencies, known as Band III, have become available for commercial radio communication systems. Two national networks were originally established and eight regional networks using the Band III frequencies. These networks have now merged. The system provides local or national coverage to organisations who would find a private commercial network for their own use prohibitively expensive to establish. These systems are known as the trunked networks and their operation and pricing is different from the normal cellular systems. All communication is restricted to 1 minute. Data can also be transmitted on the system from an office computer with an approved modem to a mobile similarly equipped or a laptop computer and printer can be used. PMR is covered in Chapter 11.

Private communication network (PCN)

It is expected that the general public will eventually have a pocket telephone which can be used both in and out of the home. Eventually systems may evolve for a single number to be called to contact a person wherever they are, through a national cellular system, but such a system will cost a great deal of money for the infrastructure as a large number of cells and microcells will be required to give satisfactory coverage for such small and less powerful phones.

A band of frequencies for a UK personal communication network has been allocated from 1710 to 1880 MHz. Systems are being installed to provide a service in this band and the Mercury system started to become operational during 1993.

Competing systems using microcells in the cellular band are being installed by existing cellular companies who will be able to operate at lower frequencies (900 MHz) and incorporate the system with the installation of the GSM digital network (see the section on compatibility).

A different approach was made by Hutchison who established a network which allowed a digital telephone to be used as a cordless telephone in the home or office and within a defined radius (approximately 200 m) of a telepoint to replace the necessity to use a phone box. Calls could only be made from a telepoint; they could not be received on this system. Hutchison installed approximately 12 000 (December 1992) high street base stations throughout the UK for their Rabbit system. A frequency of 864 MHz was used between the phone and base station. All telepoint phones sold in the UK conformed to the Common Airtime Interface (CAI – see the section on compatibility) and were able to be used on any other operator's system using this standard. Unfortunately this was not a commercial success in the UK and ceased operation in December 1993 (Chapter 13). It is still used in other countries.

Satellite systems for mobiles

Satellites provide the possibility of reaching remote areas where other methods of communication are impossible. Geostationary satellites (described in Chapter 2) always appear to be in the same position above the earth and are, therefore, used as communication satellites for broadcasting, telephony and data. Transmitters and receivers need a clear line of sight to the satellite, unobstructed by buildings, trees or hills.

Since 1965 the INTELSAT geostationary satellites, which carry approximately two-thirds of the world's telephone calls and all the transoceanic television, have increased in size, facilities, power and numbers. The system is owned by a consortium and over 100 nations have shares in the organisation. The system is operating in over 160 countries through over 700 earth stations. A geostationary satellite system

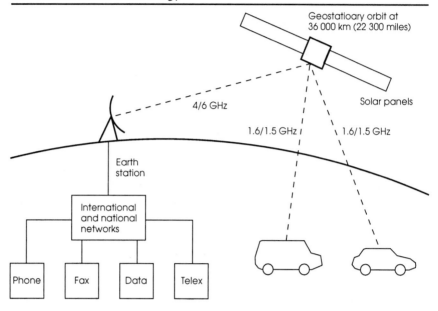

Figure 7.1   *INMARSAT system was designed as a geostationary satellite system for communication with ships at sea. Over 6000 ships already use the INMARSAT communication satellite system. The system is being extended to enable low speed data to be sent (600 bits/s) between the mobile, satellite and earth station. The system is all digital and functions as a store and forward system. Separate frequencies are used for the up and down paths to the satellite. The system is designed to allow data to be received on a non-stabilised omni-directional aerial. There is considerable redundancy in the code and the data is interleaved to allow correction should the signal experience fading during reception. INMARSAT is also used for providing some commercial aircraft with phone and fax facilities.*

known as INMARSAT (International Maritime Satellite Organisation) was developed for voice and data communication with ships and became operational in 1982. It is this system which is providing a world land mobile communication system. A European satellite, EUTELSAT, and an American satellite, AMSC, provide regional coverage. A service began in January 1991 for a slow speed data link to mobile earth terminals through the INMARSAT and EUTELSAT satellites. Both INMARSAT and AMSC can provide voice communication with the necessary equipment.

The market for satellite communication is considered to be positioning (navigation), tracking, messages and mobile telephone systems. Very accurate positioning systems such as Navstar (GPS) already exist (Chapter 2). The large market would be telephones but because of the long distance to a geostationary satellite (36 000 km) the losses are great and parabolic dishes and relatively high transmitter power are required. Powerful satellites using spot beams allow operation of receivers with omni-directional aerials but the power required from a mobile phone transmitter would make its operation dangerous.

A proposed system for satellite mobile communication by the year 2000 involves a number of small low orbit satellites circling the earth which itself is rotating. Each satellite would, therefore, scan the earth during a number of orbits. Such a system would allow a message to be stored and forwarded but it would not allow voice communication if a small number of satellites were used. This could only be achieved if a large number of satellites were employed (estimates range between 12 and 77 depending upon the orbit). A proposed orbit of 1000 km would reduce losses and the power required for transmission. The systems would be controlled from regional base stations.

Motorola are continuing to propose the Iridium system and present plans intend the system to be operational by 1998. The estimated cost is $3.37 billion (1993). It proposes 66 low earth orbit satellites and intends to offer subscribers voice, paging, facsimile and data with hand-held wireless telephones. Proposals for other systems will obviously be made and the cost of establishing the network will be considerable as will the political difficulties. Global Mobile Satellite Communications have been allocated frequencies in the S band (2.5 GHz) and the L band (1.6 GHz).

Mobile communication systems for aircraft passengers

The INMARSAT satellite is also used for providing some commercial aircraft with mobile phone and fax facilities. Passengers are able to make calls direct from their seat via cordless phones to equipment linked to the aircraft's satellite communication equipment. This links to ground earth stations via the INMARSAT satellite. In the more sophisticated systems using several channels it is possible for ground staff to monitor the technical performance of the aircraft while it is in flight.

Although satellites are the only practical way of providing telephone communication on transoceanic flights, there is a considerable number of air routes passing over land. In Europe and North America these flights far exceed the transoceanic traffic. Terrestrial systems can be provided far more cheaply than a satellite system. In North America a system (North American Terrestrial System – NATS) has been operating since 1984 and over 1700 aircraft are at present equipped to use the system. The system covers the whole of America and Southern Canada. It uses a split duplex FDM system and the frequency bands allocated are 849 to 851 MHz ground-to-air and 894 to 896 MHz air-to-ground. The Federal Communications Commission has now allocated licences to five other companies who must share the available bandwidth. It is the intention that the present system will be updated and a digital system will be installed.

In Europe, the European Telecommunications Standards Institute (ETSI) have set up a committee to produce a standard. The American terrestrial APC system (Airborne Public Correspondence), which is the generic name for these systems, has been renamed Terrestrial Flight Teiephone System (TFTS) for Europe. The frequencies allocated for TFTS are 1670 to 1675 MHz ground-to-air and 1800 to 1805 MHz air-to-ground. A digital system will be used and a mixture of frequency division and time division multiplexing (FDM/TDM) is proposed. The available bandwidth is divided into 164 radio channels in each direction with a channel spacing of approximately 30.3 kHz. Each channel has a data rate capacity of 44.2 kbits/s and is able to support 16 TDM 2.4 kbit/s user circuits plus the necessary control channels. Sixteen aircraft are, therefore, able to communicate simultaneously on one radio channel with one ground station. Initially the voice coder operates with a data rate of 9.6 kbits/s and, therefore, four 2.4 kbit/s circuits need to be combined to provide the data capacity. However, codecs are being developed to function with half and quarter the present data rate. Data and fax can operate by combining a number of user traffic circuits.

The system operates on a cellular basis with handover as the aircraft crosses cell boundaries. The radio channels are divided into three categories and allocated to three types of ground station. These are airport, intermediate or on route and they differ in the maximum altitude at which an aircraft can use them. The radiated power varies according to the type of station and these variations affect the distance before the frequencies can be reused (Chapter 10 explains reuse of frequencies in a cellular system). It is estimated that approximately 40–50 on route stations will be required for optimum coverage for cruising aircraft.

Analogue and digital systems

Initially all communication systems were analogue. However, with the growth of computer technology and the availability of dedicated integrated circuits digital technology is gradually replacing analogue in the latest systems. We originate our sound and sight information in analogue form and must convert digital signals back to an analogue sound or a visual display for humans to be able to understand it. However, for transmission purposes there are distinct advantages in sending the messages in digital form and future generations of equipment will concentrate on this format. Digital systems will be explained in more detail later in the chapter.

Compatibility

The major problem facing a user with different systems to choose from is that different equipment needs to be used for different systems due to the incompatibility of the interface standards. The next generation of equipment tries to alleviate some of these problems by producing the GSM (Groupe Speciale Mobile) specification for digital cellular mobile systems throughout Europe and DECT (Digital European Cordless Telecommunications) for European digital cordless telephony. This standard specifies the radio characteristics and communication protocols for systems capable of transmitting voice and data over a distance of approximately 200 m. A UK development of a digital specification is known as CT2 and a Common-Air-Interface (CAI) specification has been derived from it. (see Chapter 13 for details of these specifications.)

The ultimate specifications for Europe may be those of the European Commission RACE Mobile programme (Research into Advanced Communications for Europe) which is trying to integrate all the mobile services including cordless, cellular, paging and data transmission up to 2 MHz. It anticipates a mass market using a single personal communicator with intelligence and ports to support data communications and fax but be very cheap to buy. The specification requires three layers of cells of varying sizes. These are known as macrocells, microcells and picocells. It is proposed that the radio spectrum will be assigned to different services according to local need and this would provide more efficient use of the spectrum. It is intended that the RACE proposals may eventually form a world specification although Japan and America may continue with their own separate systems.

Basic system

Simplex and duplex systems

Systems can either transmit only from a base station and receive at the subscriber, as in paging systems, or transmit and receive in both directions. There are three possibilities for the latter systems which provide some operational advantages depending upon the use intended for the system.

1. A single frequency can be used for transmission and reception in both directions. This is known as single frequency simplex. It saves on frequencies but it allows conversations to be heard in full by everyone on the system. The system may require a base station to be in constant contact with a number of mobiles. A single frequency operation, however, allows the mobiles to converse with each other directly and the channel can become continually occupied with mobiles conversing and the base station may find difficulty in contacting mobiles. Only one person can use the channel at any particular time.

2. A separate frequency can be used for reception and transmission but again only one person at a time can use the system. This is known as semi-duplex. The base station transmits to all mobiles on one frequency and they respond on a different frequency. Mobiles cannot communicate with each other and only the base station can be heard.

3. Two frequencies are again used for transmission and reception but the system is designed for simultaneous working. This is known as duplex and is used for cellular mobile telephones.

Although all the radio systems previously mentioned appear to be different the requirements of the systems are very similar. All of them must have the following:

1. A radio frequency on which to transmit and receive.
2. A transmitter and aerial system which can send sufficient power into the reception area in order for the receivers to work correctly.
3. A receiver and aerial system which is sufficiently sensitive to give good reception in the required areas.
4. Battery power for the mobile system.
5. A signalling, charging and control system.

Communication frequencies

It has been shown in previous chapters that the transmission characteristics of a radio wave and its attenuation in a particular environment depend upon its frequency.

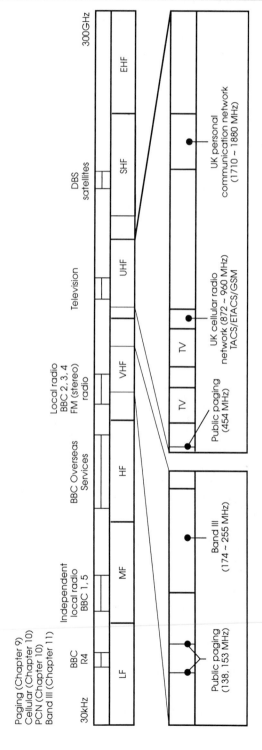

Figure 7.2 *The radio spectrum is allocated to many services. The simplified diagram for the UK shows the position of the public mobile radio systems in the VHF and UHF bands.*

The size and type of aerial and the technology used in the transmitter and receiver again depend upon the radiation frequency. Unlike a cable communication system which can be expanded as growth occurs, the radio spectrum is finite and, therefore, has to be closely controlled and allocated on an international basis in order to ensure priorities are satisfied, interference is limited and the maximum use is made of the available spectrum.

International committees

The international committee for providing the radio regulations is the International Telecommunications Union (ITU). A restructuring of this organisation is due to come into force on 1 July 1994 but, due to the importance of maintaining the ITU at the forefront of the development of world communications, it has been agreed by all the participating authorities that the structure and new working arrangements will operate from 1 March 1993.

The present structure has been in operation since 1947 when the ITU became a specialised agency of the United Nations and became responsible for the use of the radio spectrum. The ITU has its headquarters in Geneva. Periodically a World Administrative Radio Conference (WARC) has been called, whose members were responsible for the allocation of frequencies for specific purposes. These included international broadcasting frequencies, satellite frequencies and all communication systems including mobile radio requirements. For the purpose of allocating frequencies the world is divided into three regions:

Region 1 includes Europe, Africa, Russia and the Middle East.
Region 2 includes North and South America.
Region 3 includes the Far East, Japan, Australia and New Zealand.

The ITU was comprised of a permanent general secretariat, the International Frequency Registration Board and the secretariats of the two International Consultative Committees. These were CCIR (International Radio Consultative Committee) and the CCITT (International Telegraph and Telephone Consultative Committee). These committees provided the forum for reaching international agreement on standards and operational procedures.

The committee responsible for co-ordinating European Common Proposals (ECPs) for presentation at WARC was a working group of the European Radio Communications Committee of the Conference of European Posts and Telecommunications (CEPT).

New ITU structure

The new structure involves a new committee taking over the responsibilities of the CCITT and CCIR and the disappearance of these two major standards committees. Three new sections are to be set up:

1. Telecommunications Standardisation Committee. This committee is responsible for the standards formally set by the CCITT and the CCIR.

2. Radio Communication Sector. This section is responsible for the efficient management of the radio frequency spectrum both in space and for terrestrial use, formally the responsibility of the CCIR, plus the activities which were previously the responsibility of the International Frequency Registration Board (IFRB). The IFRB recorded the frequency assignments of all radio services and the orbital positions of space stations throughout the world.

3. Development Sector. This sector is relatively unchanged after a previous reorganisation in 1990. Its responsibility is the improvement of telecommunication equipment and systems in developing countries. This is achieved by advice, information and training in order to produce self-reliance for the developing countries. The Telecommunication Development Bureau (BDT) and the Centre for Telecommunication Development (CTD) were set up in 1990 to perform these duties.

The plenipotentiary conferences will be held every four years with an aim to discuss policies relating to long term issues and take decisions on strategic plans presented by the elected council. Telecommunication standardisation conferences will also be held every four years and will also operate through study groups. The conferences will discuss the work of the study groups and will approve, reject or modify their recommendations. The conferences will also approve the schedule of work for the section.

The new and old structures of the ITU are shown in Figure 7.3.

European Telecommunications Standards Institute (ETSI)

In 1988 a new committee was set up on the joint initiative of the European Commission and the European Posts and Telecommunications Conference (CEPT). Its aim is to set common telecommunication standards throughout Europe to aid European telecommunication systems and business as part of the European single common market. The committee, ETSI, is a European standards making body which drafts and publishes European Telecommunication Standards (ETSs). It also publishes Interim European Standards (I-ETs) and Technical Reports (ETRs). ETSI consists of:

1. General Assembly (governing body)
2. Technical Assembly (directs and approves the production of standards)
3. Secretariat (administration and management)
4. Technical Committee (manages the work on standards making)
5. Sub-technical committees (carry out detailed work on standards)

Membership of ETSI is open to any company or group from a country within CEPT who has an interest in European telecommunications and is prepared to pay the appropriate membership fee.

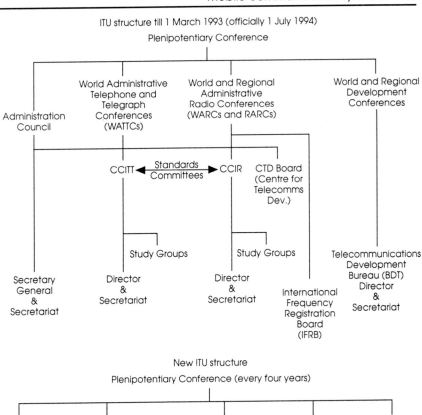

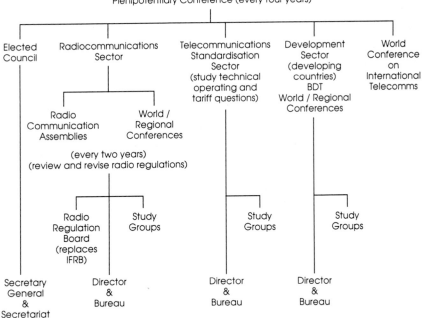

Figure 7.3 *The ITU structure.*

Radiocommunications Agency

In the UK, the Department of Trade and Industry (DTI) has the responsibility of licensing users of radio transmitters and allocating the frequencies. It is also responsible for ensuring the regulations are complied with and interference is not generated. The executive agency of the DTI which performs this work is the Radiocommunications Agency.

In the UK the primary users of the radio spectrum are as in Table 7.1.

Table 7.1

	30 kHz to 1 GHz	1 GHz to 30 GHz
Government fixed and mobile	37%	21.9%
Government and civil radar	2%	9.5%
Land mobile (civil)	8.3%	5.8%
Civil aviation	7.3%	12.7%
Broadcasting	25%	2.8%
Fixed links	11.3%	43%
Emergency services	6.5%	
Meteorology, shipping, astronomy	2.6%	4.3%
	(inc. Std. time freq.)	(inc. emergencies)

(Figures DTI 1992)

Why digital?

Speech and vision are analogue signals. The information is carried in the varying amplitudes and frequencies of the signal. Any alteration to either of these parameters distorts the signal. Irrespective of how the signals are modified for transmission they must be converted back to pressure waves by some type of speaker in order for a person to hear the message and as a light display such as a TV screen, LED display or written on paper in order for a person to see the message. Analogue transmission systems convert sound and light into electrical signals but keep the original amplitude and frequency variations throughout the system as the transmitted signal. Unfortunately all transmission systems distort the signals and in a transmission system the signal experiences losses, modulation, amplification, filtering, equalisation, phase changes, demodulation, noise, distortion, etc. Any change to the relationships of the varying amplitudes of the recovered signal to that of the original or to the frequencies which comprise the signal results in distortion. Noise is also added to the system as amplitude variations. When a signal is passed over an analogue circuit the signal continually deteriorates and the distance that can be covered depends upon the amount of distortion that can be tolerated before the recovered signal is unacceptable. The problem with an analogue signal is that its amplitude and frequency at any particular time is unknown as this is the nature of the signal and cannot, therefore, be corrected.

The digital signal

The growth of digital computers from the 1960s ensured the continuous development of components and great packing densities of active components were achieved on a single piece of silicon. Equipment became more complex but smaller as single integrated circuits replaced thousands of components. The price reduction was also considerable as the scale of production increased. Digital technology requires repetitive circuitry for its operation and computer technicians are aware of how similar many boards appear in the system. Analogue boards are custom built for a particular process and must be aligned to produce the required performance. Digital circuits are more tolerant and once the system is working usually need no further alignment.

Binary code

All information in a digital system is carried in the form of a binary code. This consists of two possible states for a pulse. It is either on (1) or off (0). A particular code then consists of a number of binary digits and the value of any particular digit is determined by its position in the code. If we have eight digits in the code (byte) it means we can have $2^8 = 256$ separate arrangements of the digits each of which can represent a particular piece of information. These codes range from 0000 0000 to 1111 1111. A greater or lesser number of digits will alter the number of possible codes and, therefore, the number of individual pieces of information that can be coded.

As all information in a digital system is composed of pulses which are either 1 or 0, providing a 1 or 0 can be recognised, irrespective of distortion, a new 1 or 0 can be regenerated anywhere in the system and the signal can be returned to its original state. This is the bonus for the transmission engineer. Signals are still distorted during transmission but they can be regenerated and the quality can be as good as the original. Digital signals have the characteristic that they are excellent while the 1 and 0 can be recognised and regenerated but suddenly become rubbish the moment the noise and distortion prevent recognition. Distance no longer determines the quality of the signal.

Analogue circuits contain components and circuits which do not allow the transmission of digital signals without either severe distortion or complete loss. If, therefore, digital data requires to be transmitted on an analogue circuit it must be made to appear to be an analogue signal which is within the bandwidth of the analogue circuit. This conversion is made by a modem (modulator/demodulator); see Chapter 12.

In addition to the quality of transmission obtained by digital transmission systems and the relative low cost of the electronics there are also other advantages. The use of integrated digital circuits has led to a reduction in the size of equipment. This is most noticeable in the size of telephone exchanges which have converted to digital exchanges System X and Y from analogue exchanges. Rooms of equipment have shrunk to a few bays and single exchanges in a linked system of exchanges are able to provide all the switching facilities.

Once all the signals are digitised computers can handle the signals. Signals can be interleaved and switched at very high speeds, codes can be modified to save bandwidth and separate circuits can be used for control and signalling.

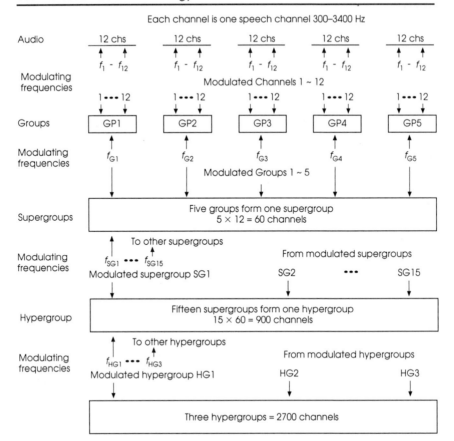

The bandwidth of each channel is 0–4 KHz. Each contains speech 300–3400 Hz. Frequency division multiplexing (FDM) allows a large number of individual channels to be combined for simultaneous transmission on a single transmission link (coaxial cable, radio link, fibre optic). Each channel eventually occupies a unique band of frequencies obtained by several modulating stages. The channel is transmitted as a single sideband suppressed carrier signal. It is recovered by reversing the process and filtering the hypergroups, supergroups, groups and channels (see Chapters 5 and 6 for modulators and demodulators).

Figure 7.4 *Frequency division multiplexing (FDM) (used on analogue circuits between cities).*

Multiplexing speech channels

Transmission engineers have always combined analogue signals in order to transmit them on a single wide-band transmission system by using frequency division multiplexing. In this method groups of 12 speech channels are modulated with individual frequency carriers and then combined and remodulated with another frequency. Five groups are combined to form a supergroup (60 channels), remodulated with a new frequency and combined with 15 supergroups to form a hypergroup (900 channels). Combinations of hypergroups and remodulation can be performed to combine 10 800 speech channels on a single transmission line or radio link providing its bandwidth is sufficient. As a result of modulating, combining and remodulating each channel has an individual carrier frequency and can be recovered by filtering the signals and reversing the process.

Digital speech channels are also combined on a single transmission line but this is performed by time division multiplexing (TDM – see later).

Digital memories

Once a signal has been digitised it can be stored in a digital memory, where it can also be processed, for as long as desired. The signals can be written and read at different speeds, time stretched, delayed, compressed and read out in a different order to that in which they were written. This has considerable applications for multiplexing and reducing or increasing data rates. It also allows complex coding functions to be performed.

Quantising the analogue signal.

When an analogue signal is converted to digital signals, pulses are produced in coded form to represent its amplitude at particular moments in time.

The number of pulses forming a particular code determines the number of amplitude samples that can be taken of the analogue signal. The number of samples must be sufficient for the listener to be unable to detect the changes as discrete jumps in amplitude. The greater the number of sampling levels the more exact is the reproduced signal. However, this requires a greater number of digits for each word in order to produce sufficient codes for the additional levels.

The analogue signal must also be sampled at a rate so that the highest frequency

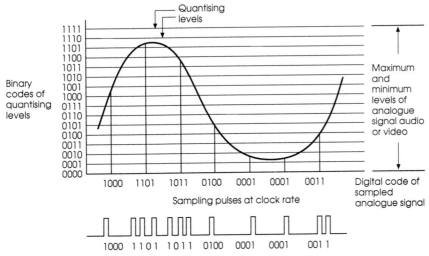

Figure 7.5 *Quantising an analogue signal. The amplitude of the analogue signal is sampled at a frequency which is normally at least twice the highest frequency contained in the analogue signal. The instantaneous level is represented by a binary code. For simplicity the figure shows 16 levels using a four-digit code. In practice an eight-digit code is used which allows 256 individual levels to be identified. Where the instantaneous amplitude does not directly correspond to a level a lower level code is produced. When reproduced this does not accurately correspond to the original signal and the difference is quantisation noise.*

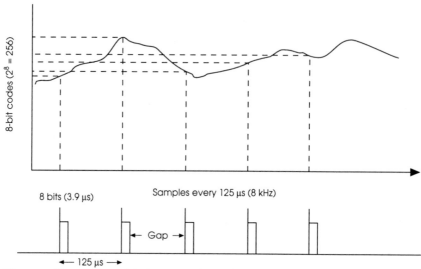

When quantising analogue telephone speech samples are taken every 125 μs. The eight pulses only take approximately 3.9 μs to transmit. This leaves a gap before the next sample.

Figure 7.6 *Quantising telephone speech.*

is sampled at least twice during a full cycle. In the BT network a telephone channel ranges from 300 to 3400 Hz (commercial speech). This means the sampling rate must be at least 6800 Hz. In practice 8000 Hz is used. Eight bits for each sample are used which provides 256 sampling levels.

The analogue signal is examined at an 8 kHz rate (every 125 micro seconds) and the 8-bit code representing the amplitude at that particular moment is read. 125 microseconds later the new code is read to represent the instantaneous amplitude. The analogue signal continues to be read and the digital coding sent to line consists of groups of eight pulses spaced at intervals of 125 microseconds. Each group of eight pulses takes approximately 3.9 microseconds to send and there is nothing between the end of these pulses and the next sample.

Time division multiplexing (TDM) – British Telecom 30-channel system

It can be seen that in the time of 125 microseconds it is possible to send 32 groups of pulses, each lasting 3.9 microseconds. This is the basis of time division multiplexing (TDM). Thirty channels are combined by taking the sample of eight pulses from ch. 1 and immediately following it with a sample from ch. 2 and then ch. 3, ch. 4, etc. Two spare channels are used for synchronisation (ch. 0 and ch. 16). It is essential that synchronisation is maintained so that when the samples are separated and recombined the correct samples go together to form the original channels.

A greater numbers of channels can be combined and sent to line by placing the 30 channels in a memory and reading them out at a faster rate. If the read-out rate is

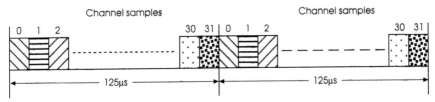

Analogue signal sampled at 8 kHz rate.
(Time = $1/f_1$ = 1/8000 = 125 µs)
Each channel sample consists of eight digits which take approximately 3.9 µs to transmit. There are 30 channels of audio and two synchronisation channels (0 and 16). To recover the analogue signal the samples must be separated and recombined as separate channels. Systems must be accurately synchronised to ensure only the correct samples of each channel are combined.

Figure 7.7 *Time division multiplexing (30 channel PCM).*

doubled, 60 channels can be combined and take the same time to transmit as the original 30 channels.

The channel bit rate for quantising a speech channel as described is:

8 bits × 8 kHz = 64 kbits; 30 channels is approximately 2000 kbits.

The maximum frequency which occurs in a digital signal is when a 1 is followed by 0. This is equivalent to one cycle. Other patterns produce a different but lower frequency. The maximum frequency of a digital signal is, therefore, half the bit rate. A single 64 kbit signal, therefore, requires a bandwidth of 32 kHz which should be compared with the 3000 Hz required for the analogue signal. This is one of the disadvantages of digital transmission systems.

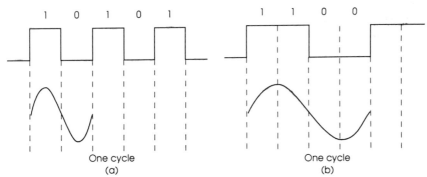

Figure 7.8 *(a) Maximum frequency with a bit pattern 1010; (b) Lower maximum frequency with a bit pattern 110011. The bit pattern determines the maximum frequency transmitted. The highest frequency occurs for a pattern of 101010. The figures show that 2 bits are equivalent to one cycle. The frequency = $\dfrac{bit\ rate.}{2}$*

Non-linear quantisation

When the signals are sampled it is possible for the signal to be between two code levels. When the signal is reformed an incorrect level is produced. This difference between the original and the reconstituted signal is noise on the system (quantisation noise).

To reduce the noise, advantage is taken of the fact that most of the analogue signal remains around the middle of the range and rarely goes to the maximum or minimum. The quantisation levels are, therefore, not linear over the full range of the scale but are spaced so that levels are placed closer together in the middle. Any differences between the signals and the levels are reduced where the problem would cause most trouble.

British Telecom network – PSTN (Public Service Telephone Network)

Cellular radio relies on the PSTN for the connection of its calls to offices and homes and is, therefore, an important part of the network. Switching to the actual radio telephone is controlled by the cellular operator's own exchanges which then feed the actual transmitters. Cellular systems are covered in detail in Chapter 10.

Telephone traffic

The telephone network works on the assumption that not everyone uses the phone at the same time. Analysis of a day's transmission shows that the use of the telephone system varies considerably throughout the day and peaks during a period known as the busy hour. The actual busy hour depends on the location of the exchange and the type of customers the exchange serves. An exchange serving mainly businesses would normally peak between 10.00 a.m. and 11.00 a.m. The amount of traffic in this period is used to decide the amount of equipment and lines used in the exchange. Sufficient equipment and lines are installed to provide a certain grade of service (GOS). The GOS determines the number of calls which cannot be connected due to the lack of equipment or lines:

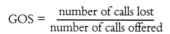

$$GOS = \frac{\text{number of calls lost}}{\text{number of calls offered}}$$

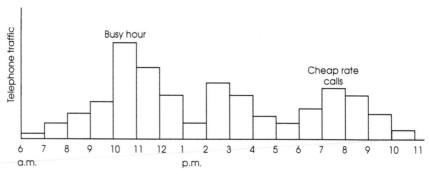

Figure 7.9 *A typical day's telephone traffic for an exchange serving a business and residential area. The pattern changes at weekends when business traffic is less. The busy hour is used for estimating the amount of equipment and lines required in order to obtain a satisfactory grade of service.*

Exchanges serving a different mix of customer such as an entertainment area, fishing industry or international calls with time differences will have a different pattern and a different busy hour.

GOS is normally given as a decimal.

For example, one call lost for every 200 calls = 0.005.

Each group of switches and lines have their own GOS and if this falls to an unacceptable figure, extra equipment or lines are installed. The overall GOS of a telephone connection is the approximate sum of all the individual grades of service in the connection.

Telephone traffic is measured in erlangs and 1 erlang equals 1 line fully occupied for 1 hour. In practice a line is not occupied for 1 hour as time is taken to set up and release a call. A single line, therefore, has less than 1 erlang capacity.

Telephone exchanges

The telephone system requires every subscriber to have a line to the local exchange, a telephone, call metering and equipment at the exchange to gain admission to the exchange routing equipment. All the rest of the exchange equipment and lines between the exchanges are shared.

British Telecom operate both digital and analogue exchanges. If it is an analogue exchange the call requires a pair of conductors to carry the call through the exchange and the call is individually switched at each stage. These exchanges normally use either switching matrixes consisting of reed relays (TXE electronic exchanges) or crossbar exchanges (TXK). Both types of exchange are known as common control as the equipment used for setting up a call is common to all customers and is only used during the setting-up procedure. The original Strowger exchanges which use individual mechanical switches and set-up equipment are now obsolete and only rarely found.

Digital exchanges are fast replacing all other types of exchanges. In a digital exchange, the incoming call, if still in an analogue format, needs to be converted to a digital format. Switching is performed by digital switches and both switches and lines carry multiplexed signals. Multiplexed signals go into the digital switches, the

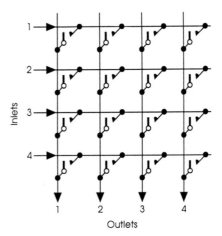

Any inlet can be connected to any outlet by closing the appropriate cross point connection. BT analogue exchanges use either reed matrixes or crossbar matrixes. Although in theory one large matrix could be used to route inlet lines to outlet lines, in a large exchange this would be an inefficient use of the matrix switches. In practice a call is routed through several small matrixes which are connected together. All the routing of a call is controlled from a single group of equipment known as common control equipment. This group of equipment handles all the switching through the exchange and is only used for the duration of setting up the call

Figure 7.10 *Principle of a matrix switch.*

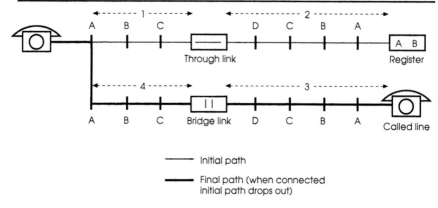

Initial path

Final path (when connected
initial path drops out)

A simplified form of an own exchange call routing on a TXE4 analogue exchange. These exchanges are used for a large number of subscribers and have a capacity of approximately 40 000 customers. A, B, C, D represent switching matrixes and in the TXE4 reed switches are used. The system is known as space switching as there must be two physical lines for each call through the exchange for voice transmission and these lines are physically switched. A number of matrixes are used to obtain the maximum efficiency of switches. The call set-up and routing is performed by common control equipment which is common to all calls and is only in the circuit while setting up and switching the call. The call is set up in two parts:

1. A calling subscriber, by lifting the phone, signals to the exchange. Common control identifies the line and selects and switches the customer to a register into which the customer dials the required number (paths 1 and 2). The coded digits are examined and the called line is examined to see if it is free.
2. If the line is free a second routing is arranged and the called subscriber is rung. When they answer, the new routing 3 and 4 is switched and a conversation can take place. In the final routing there is a bridge link which provides power to the lines, supervises the call and allows speech transmission via a transmission bridge. In the initial set-up (1 and 2) the through link is simply a connection.

   Common control must provide routing and switching within all the matrixes. In a TXE4 exchange part of common control continually scans all the lines and supervises their state (free, engaged, calling, etc.). Common control equipment is not shown here.

Figure 7.11   *Analogue telephone exchange, own exchange call.*

destination of each call is identified and the component parts of the call are time switched and separated and finally multiplexed on the output with other calls which require to go out on the same routing to other exchanges. If the call terminates at the exchange it needs to be converted into an analogue signal before routing it to the customer.

Linking exchanges

Exchanges are linked together in a hierarchical system of local exchanges (LEs), group switching centres (GSCs), district switching centres (DSCs) and main switching centres (MSCs). Subscribers are connected to a local exchange by their individual pair of wires. Groups of LEs are connected to a GSC which is connected to at least one DSC. DSCs in turn are connected to an MSC which is fully interconnected with other MSCs. Wherever traffic demands justify direct connections between exchanges they are installed, otherwise calls are routed through the different networks of exchanges.

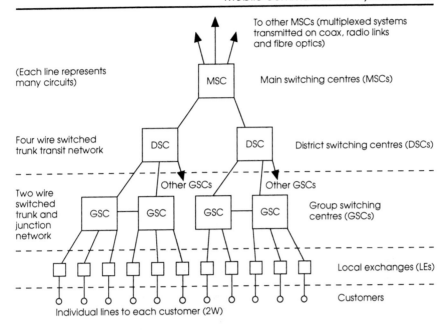

The exchanges are arranged in a hierarchical system. Whenever traffic is sufficient between two exchanges direct lines are provided.

The STD network consists of two parts:

1. This handles short distances or long distances which only require one intermediate switching centre between two GSCs. Switching is performed in the two wire part of the circuit.
2. The long distances calls requiring more than one intermediate switching centre are switched in the four wire part of the circuit. This is the transit network and high speed signalling systems are used.

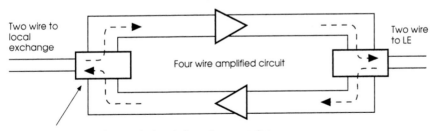

Hybrid transformers for two to four wire conversion.
Input and output of amplifiers are electrically separated

Amplifiers are used when attenuation on the cable causes too much loss of signal. Amplifiers only transmit in one direction and, therefore, amplified circuits must be four wire. Hybrid transformers prevent howl round by separating the inputs and outputs of the amplifiers.

Figure 7.12   *Interconnections of junctions, trunks and local ends to exchanges. (Analogue network)*

A digital exchange is controlled by software and the processor. Digital electronics allows a variety of exchanges to be built from standard building blocks and for a considerable number of facilities to be engineered into the system and offered to the customer. The concentration stage allows groups of customers' lines to be connected to the digital switch in the most economic way and still provide an acceptable grade of service. The main expense in a digital exchange is

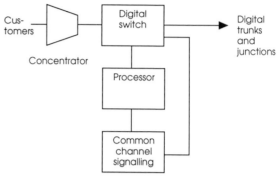

the customer's line circuit. Each customer must have one. This provides:

1. Power to subscriber's line.
2. Voltage protection.
3. Ringing current and ring trip detection.
4. Supervision of the line-on or off hook and dialling.
5. Encoding/decoding-conversion of analogue signal into digital and the reverse.
6. Hybrid 2–4 wire conversion. The customer has two wires but within a digital exchange it is a four-wire system as there are two circuits one for transmission and one for reception.
7. Testing – it is a requirement to be able to test both the line and the equipment.

Local exchanges may only contain line circuits and a concentration stage. The signals are converted to PCM and transmitted to group switching centres in a TDM format where all switching is performed.

While there are still analogue exchanges a digital exchange may be complicated by the necessity to provide both analogue signalling and common channel signalling. Additional equipment and programming must be incorporated to provide these facilities.

Figure 7.13   *A simplified diagram of an exchange which only handles digital PCM signals. Signalling for PCM is carried on a common channel separate from the digitised audio.*

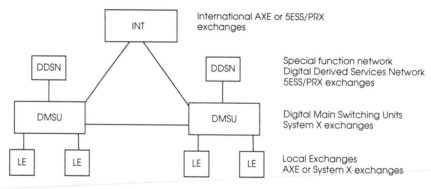

AXE manufacturer, Ericssons;
5ESS/PRX manufacturer, AT & T/Philips;
Systems X manufacturer, GEC/Plessey.

In the UK the mobile switching network uses different exchanges controlled and operated by the mobile network providers. In other countries the networks may be arranged to handle all functions.

Figure 7.14   *BT digital exchange hierarchical network.*

(a) A simplified version of a digital switch: 30 channel TDM signals enter and are separated by timing signals into separate channels. The space section of the switch then switches them into the new outgoing channels. The

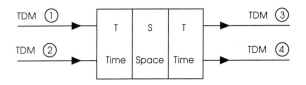

time part of the switch on the output reorganises the outgoing lines into 30 channel PCM for onward transmission.

(b) ABCD are channels in a 30 channel PCM system (1); EFGH are channels in a 30 channel PCM system (2). On the output of the switch the channels have to be reformed into new 30 channel PCM combinations for onward transmission.

Time shifting is performed by the use of digital stores. The individual incoming PCM words are written into addresses in stores. They can then be read out in any order and reformed into new PCM groups of channels.

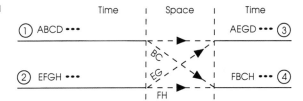

Figure 7.15 *Digital switch.*

Subscriber trunk dialling

Subscriber trunk dialling (STD) allows a subscriber to dial anywhere in the UK. To provide this facility linked numbering schemes are used whereby a group of local exchanges and the associated GSC all have the same STD code and a common charge. A subscriber dialling into an area is routed from their own area via the trunk routes, which can be radio, cable or fibre optic or a combination, to the local GSC. At the GSC the first numbers of the subscriber's telephone number identify the local exchange to which it is routed and the final numbers identify the individual subscriber.

When making a call the subscriber, by dialling a first number 0, signifies an STD call is required. The call is immediately routed from the local exchange to the GSC where the STD equipment is situated. The remaining STD destination code is dialled and the STD equipment translates these numbers into a routing code and sets up the call through all the exchanges on the route. Long distance calls are multiplexed together when carried on the trunk network. The trunk network is now almost fully digital.

Calls within a linked group are local calls and only the subscriber's personal number needs to be dialled irrespective of which exchange is dialled within the group.

The number of subscribers and, therefore, the size of any linked system is determined by the number of figures in the subscriber's telephone number. In theory two digits = 100, three digits =1000, four digits = 10 000, five digits = 100 000 and six digits = 1 000 000. Because some numbers have to be reserved for special routings the maximum number of possible subscribers is less than the maximum number of possible telephone numbers for any linked group.

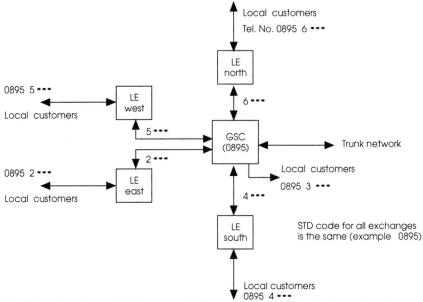

The STD code routes a call to the group switching centre. The individual customer's number identifies the local exchange by the first digits. Customers within a linked numbering scheme can call each other by dialling the customer's number. No STD code is required.

When making an STD call from a local exchange the first dialled number '0' routes a subscriber on a junction to the GSC and the STD routing equipment. The STD code dialled is then translated to a routing code which switches the circuits in the intermediate exchanges to the distant GSC.

Figure 7.16   *Linked numbering scheme.*

Director systems

Some cities have very large populations and a different system is used. This is a director system. Each subscriber has a seven digit number. The first three digits identify the exchange in the director system and the last four identify the individual subscriber. In theory this allows 10 million numbers on a director system (1000 exchanges and 10 000 subscribers on each exchange). In practice again certain numbers are used for special purposes (e.g. 0 STD, 1 services) and the number is reduced to 8 million. To enter a director system from outside its area an STD code must be dialled. There are six cities with director systems in the UK.

The director system functions similarly to STD as the exchange number is translated into a routing code by the system and this code directs the call to the called exchange where the subscriber's number identifies the required subscriber.

Digital data networks

In addition to the analogue network, for which modems are required in order to transmit data, there is also a variety of digital networks capable of various transmission speeds. When using modems a variety of techniques are used to modify the data waveform of the various data rates so that the required bandwidth does not exceed the bandwidth of commercial speech. The techniques used are frequency shift keying (FSK) for data rates up to 1200 bits/s and differential phase modulation for 2400 bits/s. These systems are described in Chapter 12. In order to transmit higher rates the analogue components such as amplifiers and transmission bridges must be removed from the circuits and they must become digital circuits. Regenerators are used to produce new pulses when they begin to deteriorate due to losses and noise. BT provide digital circuits capable of different data rates and these include Kilostream, Megastream and Satstream for satellite systems. Satstream provides facilities for remote communication and for linking international offices of multinational companies.

Kilostream provides rates of:

2.4 kbits/s for such uses as transmission between teletypewriters, VDUs, electronic mail.

4.8 kbits/s for slow scan TV and verification of credit cards.

9.6 kbits/s for high speed fax.

48 kbits/s for high speed computer transmission.

64 kbits/s digitised commercial speech.

These are only examples and once any signal is digitised it can be transmitted as the system is transparent and unaware of what the data represents.

Megastream is capable of transmitting data at rates of 2 Mbits/s (30 speech channels TDM) and 8 Mbits/s.

In addition other data services are provided and these include:

1. Circuit Switching Network: whereby a circuit can be dialled and connected via the PSTN and data terminals when required. These circuits are obviously not optimised between the users and may not be available when required.

2. Message Switching Network: whereby messages are sent together with the destination address and stored until they can be forwarded to the receiving terminal. The transmission process allows computers operating at different rates to communicate and several terminals can receive the message. Maximum use can be made of the circuits and costs can be reduced by transmission at times when the cost is least if required.

3. Packet Switched Digital Network (PSDN): whereby customers are linked by an individual network and each has an address. Data is sent to line not as a continuous message but as packages which are integrated with all the other packages as a space occurs. Because each packet is small, delays are minimal and it is possible for computers to be interactive. The packages are produced and collected and assembled by computers at each sender and receiver and are only received at the correct address. The computers are known as PAD (Packet Assemblers/Disassemblers). The systems can be linked nationally and internationally by BT.

A packet comprises 128 octets of the customer's data (Datagram). It contains the destination address, information relating to the packet position in the complete message, control and error detection data.

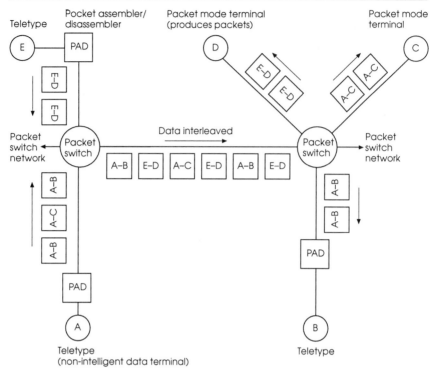

Packets of information consisting of 128 octets of customer data (datagram) are addressed and sent to line. These are switched and reassembled by either a packet mode terminal or packet assembler/disassembler.

Figure 7.17   *Packet switching.*

LANs and WANs (Local and Wide Area Networks)

Computer facilities can be very expensive but become more economic if the facilities such as printers and data storage can be shared by a number of stations. The same information is also often required by several departments for reference and updating. Networks, therefore, which allow access to the computer facilities from different access points have economic and operational advantages to the users. Networks covering short distances of less than 1 km, and often in the same building, are known as LANs (see Chapter 9) whereas networks which allow the sharing of facilities over hundreds of kilometres are known as WANs. WANs can use either the ordinary PSTN and modems or the digital data circuits previously described when high data rates and reliable service are required. As cities become cabled for TV the cable companies are also providing a separate digital telephone service as an additional facility. This service can be connected to both the BT and Mercury networks and eventually a wide network of data circuits will be spread across the cities. These networks are known as MANs (Metropolitan Area Networks). The data service, as already mentioned, can be extended to mobile radio where office computers can be accessed and information passed in both directions (Chapter 12).

Broadband transmission

Broadband networks have very wide bandwidths and are capable of carrying very high bit rates. Such networks are essential for the transmission of digital video if normal television pictures are transmitted (up to 140 Mbits/s). However, a broadband spectrum can be mutiplexed to enable different channels with different bandwidths to be used simultaneously on the network. Several permanent data channels can be multiplexed using a frequency division arrangement (FDM techniques) or a free data channel can be accessed by a frequency agile modulator which can scan and find a free channel to transmit data for the duration required. High data rates can transmit a vast amount of information very quickly and only users with such an amount of data can justify the cost of these circuits.

Considerable development on an Integrated Broadband Communication (IBC) network is being performed as part of the European RACE programme and part of the RACE Mobile Telecommunications project is to integrate this work to cover mobile communications. It is the intention that mobile broadband should operate in all environments and with data rates up to 155 Mbits/s. The allocated frequencies are within the millimetre band at 54.25–66 GHz and specifically identify 62–63 GHz and 65–66 GHz. The system will operate on a cell basis whose size will be approximately 150 m for circular cells and 350 m for elongated cells. High speed mobiles will require a system of handover between cells although it is anticipated that this will not be a universal requirement.

Integrated Services Digital Network – ISDN

Analogue electrical signals used for transmission take different forms depending upon the original analogue source of the signal. Telephone, music and video all require different analogue circuits for transmission and the transmission of high speed data requires digital circuits. A business customer uses a wide variety of equipment which may include computers, VDUs, remote terminals, phones, packet switching, video, telex and facsimile, each of which requires separate types of transmission circuits and interfacing due to the different types of electrical signals transmitted on the line.

The advantage of digitising the signals is that their characteristics are all the same and the important difference becomes the data rate at which the pulses are produced. Providing a circuit is able to transmit the data rate, a common transmission line can be used and the digital exchanges and digital trunk routes can transmit the data unaware of what it represents.

As the main BT and Mercury trunk routes are almost all digital and digital exchanges are rapidly replacing the analogue exchanges, a network can be produced which allows all types of signals to be transmitted when digitised. To gain access to the ISDN requires the local ends between the subscriber and the local exchange to be a digital link and at a basic level this can be provided on the two wire local distribution network. The subscriber will have a number of digital channels available for 64 kbit/s voice or fast data transmission. Users requiring the transmission of large amounts of data will have other types of circuits such as fibre optics.

In addition to the physical network, an internationally agreed set of standards for transmission, signalling and interfacing in order to obtain a comprehensive end-to-end digital network is required. Unfortunately many standards exist in the field at present and a universal standard may be difficult to achieve in the near future.

Seven Layer Open System Interconnect (OSI)

This system has been devised in order that development engineers are able to standardise the various parts of data equipment so that they can be interconnected and function together as a system irrespective of which company manufactured the equipment.

The characteristics of the equipment have been defined in seven layers and each layer relates to a specific function or electrical or mechanical feature. Providing the standards at each stage are adopted, equipment will connect and function as a system. In order to understand the requirements of the ISO seven level model it is necessary to make a distinction between the communication functions and the applications that they support.

The various OSI layers are as follows.

This system allows the physical and logical functions of a system to be conveniently specified. The concept is important to design engineers for clarifying specifications for different equipment which must interconnect and function in a complete system.

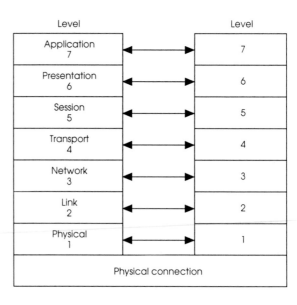

Figure 7.18   *Open system interconnection ISO Seven Level Model.*

Layer 1. Physical
This relates to the actual mechanical and electrical components used in the equipment to enable the signal to be transferred over a physical connection. It ensures all connectors are compatible, and specifications are provided for transmission codes, transmission rates and signal voltages to enable the transfer of a stream of data over a single connection path. No significance is given to any single digit at this stage.

Layer 2. Link
This relates to the circuitry which enables the data to be moved between parts of the system directly connected together. One of its main functions is to provide error detection and correction facilities. In order to accomplish this the data may have to be divided into blocks to identify the field over which the error detection code functions.

Layer 3. Network
This relates to the equipment which forms the interface between data which is not in packets and a system which operates with packets. Its function is to arrange packets and reassemble messages and control the flow of packets in a network. Data networks consist of a number of nodes interconnected by various data links and addresses are required to specify the links to be used to interconnect the required terminal node. Logical paths rather than physical paths may be specified as the same physical path may not be used for successive packets of the same message.

Layer 4. Transport
This relates to the method by which the message is moved across a system from the originating source to the destination. It must take into consideration the nature of the terminal equipment and select a network link which is capable of operating at the required data rate and quality. This relieves the terminal user of the responsibility of data transfer through the network.

Layer 5. Session
This layer is responsible for establishing, maintaining and terminating end-to-end data links.

Layer 6. Presentation
This layer provides a common language and edits, translates and maps the data into an acceptable form. The network, therefore, becomes machine independent and is not concerned with the different codes used for different tasks.

Layer 7. Application
This relates to the actual communication of the data message between the data terminals and defines the nature of the task to be performed. It provides the required user information processing functions and programs for applications such as validity checking, graphics, electronic mail, airline booking, word processing, etc., necessary for the message transmission.

The specifications for layers 1, 2 and 3 are mostly obtained by the design of the hardware while the layers 4-7 require specialised software in order to implement the specifications. Layers 1-3 are concerned with communication through the system while 5-7 are more task oriented.

The principle is that any interconnected system uses the same standards at each level and each layer can be defined independently of the other layers. Messages commence at level 7 and go down through the layers to level 1. At each stage a header is added and it is processed as required. At the receiving end the message is received at level 1 and is passed up through the different layers where the headers are removed to be finally processed by level 7.

The problem remains that many of the mainframe manufacturers already have similar but competing layered protocols which for commercial advantages they may keep, although more equipment is being produced which is OSI compatible.

8 Aerials

Irrespective of the type of radio transmitter and receiver used in an installation the important connection between the two is the transmitting and receiving aerials and the environment in between. No system can perform to its optimum specification unless the aerials are of the correct size, the position of both the aerials for transmitting and receiving are correctly located and the environment and distance between the two sites is suitable for the frequency and power transmitted.

In broadcasting we have a transmitter on one site and many receivers. In communication systems both stations are usually receiving and transmitting to each other and the aerial serves a dual purpose. In mobile radio systems the designer is at an immediate disadvantage as the mobile transmitter/receiver is in a constantly changing environment. However, the installer of an aerial for a mobile system on a vehicle can considerably affect the performance of the system by the choice of position and method of connection of the aerial to the vehicle.

The aerials for a base station transmitter/receiver are static and, therefore, the system designer has control over their position and can site them to give optimum coverage in the area required. This may require several transmitting sites to overcome environmental features such as hills and buildings and methods of ensuring that the power is only radiated in directions where it is required. In such systems a potential problem can arise when a mobile radio is able to receive the same signal from two transmitters as the paths and, therefore, the timing will be different for the separate signals. This can cause distortion. The same effect arises when radio signals are reflected from buildings and a mobile radio receives signals which have travelled over different paths. It is the responsibility of the system design engineer to anticipate these problems and extensive surveys and tests are performed before installations are made (see Chapter 2).

Transmission lines

At high frequencies a cable behaves in a complex manner and has the characteristics of a low pass filter whose cut-off frequency is determined by its components.

The cable can be represented by the primary coefficients of resistance, inductance, capacitance and leakance (insulation resistance). The cable can be regarded as many circuits in series evenly distributed along the line. The values of the components are normally given in cable specifications for 1 kilometre and as the values vary with frequency this must also be quoted. The values of the component parts of the cable are determined by the construction and the materials used.

The primary coefficients provide the secondary coefficients of the cable which determine the signal performance. The secondary coefficients are: characteristic impedance, attenuation coefficient, phase change coefficient and velocity of propagation.

When the input impedance of a very small piece of cable is measured its value is determined by the cable components and the termination. As the cable length increases, the input impedance is determined by the primary coefficients and the input signal source has no indication of the value of the termination. The input impedance of an infinite length of transmission line is by definition the characteristic impedance of the cable and signal sources and loads must be matched to this impedance (Chapter 4). In mobile radio this value is arranged to be 50 ohms as a standard.

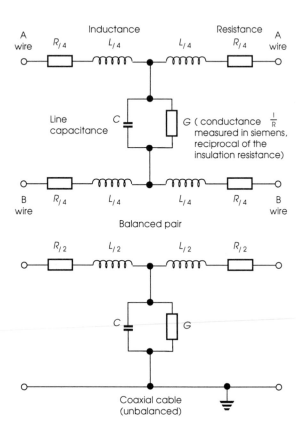

Figure 8.1 *Transmission lines. A cable can be considered as many series circuits formed from the primary coefficients of the cable.*

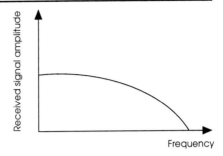

Figure 8.2 *A cable has the characteristics of a low pass filter. Losses increase with cable length.*

Standing waves

When an AC waveform is propagated along a transmission line from a source which is correctly matched to the input impedance of the cable, power is applied which must dissipate in a resistance at the termination equivalent to the characteristic impedance of the cable. If the cable is not correctly terminated then the unused power is reflected back down the cable and standing waves are created. Because it is an AC waveform the value of the voltage and current waveform at any particular point on the cable is related to the wavelength of the signal.

The two extremes are:

1. When the termination is an open circuit.
2. When the termination is a short circuit.

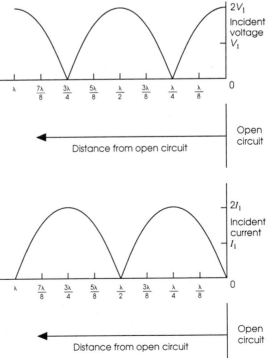

Figure 8.3 *Standing waves. The diagrams show the waveforms produced by an open circuit. If the termination is a short circuit the waveforms for current and voltage are reversed.*

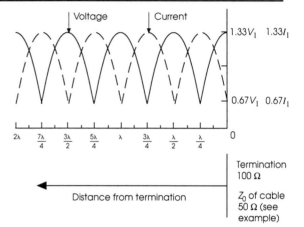

Figure 8.4 *Standing wave produced by incorrect termination.*

Termination
100 Ω

Z_0 of cable
50 Ω (see example)

At an open circuit termination the voltage is a maximum and the current is zero. When the termination is a short circuit the current is a maximum and the voltage is zero at the termination. When a resistance equal to the characteristic impedance of the cable terminates the cable, all the power is absorbed and there are no standing waves. For other values of termination the waveforms vary between the extremes of open and short circuit.

At the open circuit termination the value of the voltage is the vector sum of the incident and the reflected voltages and, as they are in phase, this equals twice the incident voltage (assuming a loss free line). The incident and reflected currents are in anti-phase and cancel. Other terminations produce lower values of reflections and standing waves. The reflected and incident waves produce a maximum voltage and current when in phase and a minimum when the phases are opposite.

Any mismatched low loss transmission line has a voltage standing wave ratio (VSWR). This is the ratio of the maximum voltage on the line to the minimum:

$$\text{Standing wave ratio} = \frac{V_{max}}{V_{min}} \tag{1}$$

The voltage reflection coefficient (ρ_v) is determined by the values of the line and termination impedances:

$$\rho_v = \frac{Z_L - Z_O}{Z_L + Z_O} \tag{2}$$

where Z_L = load impedance
and Z_O = characteristic impedance of cable

$$\text{Standing wave ratio} = \frac{V_I + \rho V_I}{V_I - \rho V_I} \quad \text{where } V_I = \text{incident voltage}$$

$$= \frac{V_I(1 + \rho_v)}{V_I(1 - \rho_v)}$$

$$\text{Standing wave ratio} = \frac{1 + \rho_v \text{ (max)}}{1 - \rho_v \text{ (min)}} \qquad (3)$$

Example (Figure 8.4)

The characteristic impedance of the cable is 50 ohms and the termination is 100 ohms:

$$\text{The reflection coefficient} = \frac{100 - 50}{100 + 50} = \frac{50}{150}$$

$$= 0.33 \angle 0°$$

The maximum voltage is therefore $1.33\,V_I$ (from formulae 1, 2 and 3)
The minimum voltage is therefore $0.67\,V_I$.

The resonant aerial

Dipole

In order to obtain the optimum operating efficiency of an aerial its size must be related to the operating frequency. When the transmitting aerial is a half wavelength of the operating frequency then a considerable amount of RF energy is radiated. In the low and medium frequencies wavelengths are very long. For practical and economic reasons, therefore, short transmitting aerials are used which are a small fraction of the wavelength. The aerial is mounted above the earth and the power is fed between the earth and the base of the aerial. At the high frequencies used for mobile radio the wavelength is short and size is not a problem.

The theory of radiation is given in Chapter 2. As the ends of the half wavelength aerial are open circuit there will be no current at these points but maximum voltage. As the signal is an AC waveform, one-quarter of a wavelength back from the open circuit the voltage is a minimum and the current is a maximum. The impedance of the aerial (voltage/current), therefore, varies throughout its length. If we centre-feed the aerial the input impedance is low whereas, if the aerial is end fed, the input impedance is high. In a practical aerial with losses the impedance at the centre is approximately 73 ohms and this is a good match for a coaxial feeder. The importance of matching impedances for maximum power transfer has been

When a λ/4 transmission line is made an open circuit a standing wave is produced. As impedance $Z = V/I$ it can be seen that the impedance changes along the length of the transmission line.

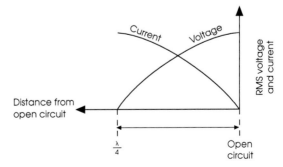

Figure 8.5 *A quarter wavelength transmission line.*

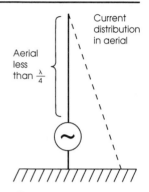

Figure 8.6   *At low and medium wave frequencies wavelengths are very long and an aerial can only be a small part of a wavelength.*

described in Chapter 4 and, therefore, for maximum power to the aerial it has to be matched to the transmission line which feeds it. If a two wire feeder is used, which has an impedance measured in hundreds of ohms, it must be fed to a part of the aerial which matches its impedance rather than to the centre. There are advantages and disadvantages to both coaxial and two wire feeders which are both technical and operational and often a radio station will use a combination. However, at high frequencies the losses on a two wire feeder are much greater and would not be used. The input impedance of an end fed dipole is approximately 3600 ohms.

If the dipole is centre fed the input impedance is low (high current, low voltage). In a practical aerial this is approximately 73 ohms. A half wave dipole is an efficient radiator of RF energy.

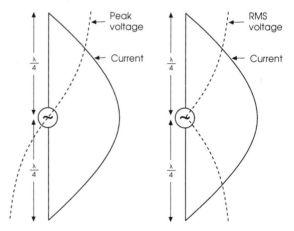

Figure 8.7   *Voltage and current distribution in a half wave dipole.*

Folded dipole

The impedance positions associated with a dipole can be changed by making a folded half wave dipole. If the half wave dipole is terminated with a short circuit its ends will have zero voltage and the current will be maximum. This is the opposite of the situation existing with the dipole and the centre is now high impedance. The value is approximately 292 ohms. This impedance can be reduced to approximately

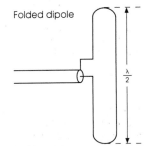

Folded dipole

Figure 8.8 *The folded dipole is a λ/2 dipole with a short circuit across its ends. This causes the centre feed input impedance to be approximately 292 ohms. This can be reduced with directors and reflectors or the use of baluns.*

75 ohms by the addition of parasitic elements (reflectors or directors – see the section on the Yagi antenna) or a balun can be used to provide the match with the feeder. Baluns are balanced to unbalanced transformers whereby the balanced aerial about earth is matched to the unbalanced feeder. Chapter 4 shows how transformers can change impedances. Conventional wire wound transformers can be used for frequencies below 30 MHz and power below 5 kW. Above these limits the balun employs transmission line techniques whereby the use of quarter wavelength transmission lines can be connected in such a way that impedance changes and balanced to unbalanced arrangements can be achieved.

In mobile radio the standard impedance is 50 ohms and both end fed dipoles and centre fed folded dipoles are used with matching devices for both mobiles and base stations. These will be described in more detail.

The length of the dipole corresponds to a half wavelength of a particular frequency. Above or below this frequency the aerial has an inductive or capacitive

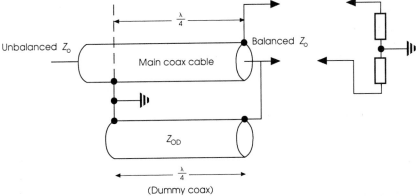

(Dummy coax)

Quarter wavelength transmission lines can be used to produce a 1:1 impedance ratio balun.

The balanced circuit is connected to the outer screens of the two coaxial lines. As these are λ/4 with a short circuit at the unbalanced end they appear as an open circuit at the balanced end and the earth is effectively not connected to the balanced circuit. The screen is also connected to the inner of the main coaxial cable and, therefore, at the unbalanced end the signal is on the inner and outer of the coaxial cable. There are many circuits which use transmission lines for transformer purposes at high frequencies. Specialist books should be consulted for further study. The principles are the same as described but details change depending upon the application.

Figure 8.9 *Balun – balanced to unbalanced matching.*

(a) If the load Z_L is different to the characteristic impedance of the transmission line, a quarter wavelength of transmission line of different Z_o can be inserted between them and a match obtained:

$$Z_{o2} = \sqrt{Z_{o1} \times Z_L}$$

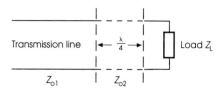

(b) Method using $\lambda/4$ short circuited transmission line. A short circuited $\lambda/4$ transmission line is placed across the load. The short circuit appears as an open circuit across the load due to the $\lambda/4$ transmission line impedance change. The impedance of the $\lambda/4$ transmission line varies from short circuit to open circuit along its length and a point is found to match it to the main transmission line.

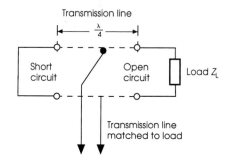

Figure 8.10 *λ/4 transmission lines to match unequal impedances.*

component and is resistive only when the aerial is resonant. In a practical aerial the length has to be slightly less than the half wavelength of the frequency because losses affect the voltage and current distribution.

Radiation

The direction in which an aerial radiates or receives is of major importance when designing systems. This feature is affected by the construction of the antenna. A dipole radiates equally in all directions when mounted in free space conditions. This is known as omni-directional and is used as the reference (0 dB gain) against which the gain of directional antennae is compared.

Yagi transmitting and receiving aerial

The Yagi aerial is familiar to everyone as it is the common aerial used for the reception of television on the outside of houses. It is also extensively used for VHF and UHF transmission and reception. It consists usually of a folded dipole radiating element with a number of parasitic elements. The size of the aerial is directly related

The dipole radiates an omni–directional pattern. This pattern is normally modified in base stations in order to direct the radiated energy to particular areas. Aerials for mobiles, however, have to receive and transmit in all directions and are omni–directional.

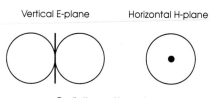

Figure 8.11 *Half wave dipole polar diagrams.*

Radiation pattern of a vertical dipole

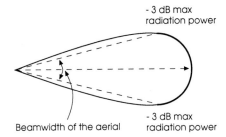

Figure 8.12 *The beamwidth is the angle subtended by the –3 dB points on the aerial radiation pattern (0.707E$_{max}$ – radiation has fallen to half its maximum power).*

Directional aerials are produced to radiate in a particular direction (forward lobe) but produce an unwanted additional back lobe. The front-to-back ratio E_F/E_B is obtained by measuring the field strength of E_F and E_B at the same distances from the aerial.

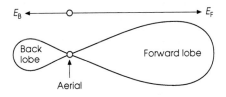

Figure 8.13 *Front-to-back ratio of an aerial.*

to the transmission frequency and consequently VHF aerials with many directors can place a large loading on a mast. The directors not only increase the gain of the aerial but reduce the beamwidth and, therefore, in mobile systems a compromise with all these factors has to be reached. The number of elements is usually between four and five.

The spacing of the elements on a Yagi aerial are critical and both reflector and directors are used to increase the radiated signal. The reflector is the element mounted behind the dipole and has induced in it a lagging voltage by the radiating dipole. This causes a lagging current to flow in the reflector and the reflector to radiate. The spacing between the dipole and the reflector is approximately 0.25 of the wavelength and

The directors and reflector absorb power and reradiate it. The phasing of their signals is arranged to add to the dipole signal in a forward direction and subtract from signals in a backward direction. In order to increase the interaction between the elements and to conserve space, the elements are placed close to the dipole. The necessary phase differences are obtained not only by the spacing but also by making the reflector inductive and the

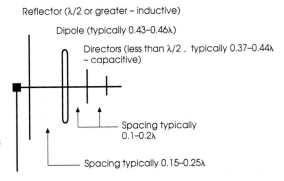

directors capacitive. This is achieved by altering the size of the elements in relation to a wavelength of signal. The relationship changes for different frequencies. Adjustment of the spacing effects the front-to-back ratio and side lobes. Calculations can be complex for size and spacing and experiments are often used. Typical gains are: three element, 4–5 dB; four element, 6–7 dB; six element, +10 dB.

Figure 8.14 *YAGI aerial array.*

it is also approximately 5% longer than the dipole. The radiation from the dipole and the reflector tend to cancel in the backward direction but add in a forward direction. Cancellation is not complete and Yagi aerials do transmit power in a backward direction. This is measured in decibels and given in manufacturers' specifications as the 'front-to-back ratio' which relates the power transmitted in both directions as a ratio.

The dipole also induces a leading voltage in the directors which causes a leading current and radiation from the directors. This radiation reinforces the dipole forward radiation but cancels any reverse radiation. However, this effect only occurs for specific dimensions and spacing related to a specific frequency. Practical aerials place the reflector 0.15 to 0.25 of the wavelength behind the dipole and directors between 0.1 and 0.2 of the wavelength in front of the dipole.

Aerials with directors and reflectors are selected for specific bands of frequencies according to their dimensions and spacing in order to obtain the desired radiation.

The placing of elements in front and behind the folded dipole modifies the centre input impedance. If a resonant dipole is used its input impedance is reduced from 73 ohms to approximately 50 ohms by the addition of a reflector and a director and is reduced much further by additional directors. This low impedance causes mismatch and standing waves on the 50 or 75 ohm feeder. By using a folded dipole, whose initial input impedance is 292 ohms, parasitic elements can be added to produce the required impedance.

Bandwidth of an aerial

This is the band of frequencies for which a particular aerial is considered satisfactory. Because the dimensions of length and spacing are chosen for a particular frequency the performance is only satisfactory within a confined bandwidth. The bandwidth is defined as the band of frequencies for which the radiated power in the required direction is less than 3 dB down on the maximum radiated power. The diameter of the elements affects the bandwidth and this can be increased by increasing the diameter of the elements from which the Yagi antenna is constructed.

UHF aerials

The dimensions of UHF aerials are much smaller than VHF and this provides advantages in construction. One problem suffered by UHF is due to rain hitting the antenna and inducing noise into the system. Fibre glass radomes can be used to house the UHF aerial and prevent contact with the rain. In addition, printed circuits can be used for the elements and a large number of directors can be provided and housed in radomes whose length is only a little over a metre.

Microwave dishes

Parabolic dishes are used for the reception and transmission of microwaves to satellites and terrestrial communication links. These dishes receive the radio waves and

Figure 8.15 *(a) A British Telecom relay station. The receiving dishes are aligned with transmitter dishes for line-of-sight transmissions. The signals are remodulated onto new carriers and retransmitted to the next receiving station. The distance between stations is normally about 25 miles and longer distances use multi-hop links.*

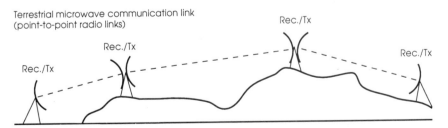

Figure 8.15 *(b) Multi-Hop radio link using microwave frequencies. Receivers and transmitters must be in line of sight and the dishes accurately aligned. When the station receives and transmits, the frequency is changed before retransmission. This avoids interference should the receiver receive signals from a previous transmitter in addition to the one directly transmitting to it.*

focus them through the parabolic focal point where the receiving antenna is placed. This concentrates the received signal through this particular point for maximum signal reception. For transmission, the transmitting antenna is placed at the focal point and this concentrates the transmitted radio waves from the parabolic reflector into a parallel beam of rays.

Radio waves at these frequencies travel in a line of sight and must not be obstructed. Due to the curvature of the earth terrestrial transmission is usually a maximum of 25 miles and multi-hop links with transmitters and receivers are used to cover longer distances (the London to Birmingham BT radio link uses four hops).

Receiving and transmitting dishes must be accurately aligned for maximum signal reception. Under certain atmospheric and geographical conditions, especially when the beam passes over water, the beam can be bent from the actual horizontal line-of-sight path. (See Chapter 2 for diagrams of various dishes and explanations.)

Aerial gain and effective radiated power

The gain of an antenna is obtained by concentrating the radiated power into a narrow beam. A comparison is made with either a dipole or an isotropic antenna (radiates the same in all directions and is a theoretical antenna – the gain of a half wave dipole compared to an isotropic radiator is 2.15 dB or 1.64). The aerial gain is defined as the ratio of the two powers required to be transmitted by the reference and compared aerial in order to receive the same signal strength at some distant point in a direction of maximum radiated signal.

This means that a directional aerial radiates more effective power than the isotropic radiator. The multiplication of effective radiated power from the aerial is the power ratio gain of the aerial. If for example an aerial has a gain of 20 dB, this would be a power ratio of:

$$
\begin{aligned}
\text{dB} &= 10 \log \text{power ratio (Chapter 3)} \\
20/10 &= \log \text{power ratio} \\
\text{antilog } 20/10 &= \text{power ratio} \\
\text{antilog } 2 &= \text{power ratio} \\
100 &= \text{power ratio}
\end{aligned}
$$

If the aerial radiates 10 watts, its effective radiated power is 1000 watts.

Stacking and baying

When high gains and narrow beamwidths are required antennae can be stacked above each other with a space of 1 to 2.5 wavelengths between each aerial depending upon the frequencies used. Two aerials would effectively double the power (3 dB) and, therefore, the gain of the individual aerial would be increased by 3 dB. A practical system provides slightly less gain than in theory.

A similar gain can be obtained by installing the aerials side by side with again dis-

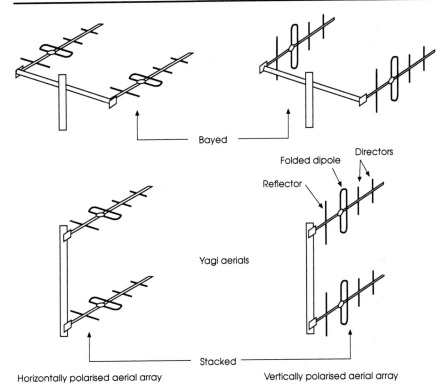

Figure 8.16 *Baying and stacking arrangements increase the gain of the aerial array and reduce the beamwidth (baying reduces horizontal beamwidth, stacking reduces vertical beamwidth).*

tances of 1 to 2.5 wavelengths between the aerials depending upon the frequencies used. This is known as baying. Both systems of installation can be combined with four antennae being used in a 'stacking and baying' arrangement.

Although both systems increase the gain, the beamwidth is reduced. Stacking halves the vertical beamwidth and baying halves the horizontal beamwidth. This can be an advantage in some systems. It is essential with such systems that the aerials radiate in phase otherwise the beam is skewed. This again could be used to advantage should an interfering signal be in line with a normal transmission. Special coax phasing harnesses are designed to feed these arrangements of aerials.

When the aerials are installed on masts the radiation pattern is affected by the mast itself. The spacing between the aerials and mast as a relationship to radiated wavelength considerably affects the polar diagram and must be taken into consideration when designing a system.

Collinear antenna

The Yagi antenna produces gain with directors and reflectors and these change the pattern of the radiated energy in order to concentrate it into a narrow beam. Stacking and baying also increase the gain and the radiation patterns are determined by the spacing and phasing of the aerials.

Omni-directional aerials can be used in similar arrangements in order to provide gain and to alter their pattern of radiation. If two half wave dipole antennae are installed one above the other with the correct spacing and the signals are fed in phase then the increase in gain can be 3 dB. The radiated beam is narrowed in a vertical plane although the horizontal plane remains omni-directional. Increasing the number of dipoles increases the gain and four produce a 6 dB gain. Special coax harnesses are used to ensure that the phase of the signal is exactly the same to each parallel fed antennae. As the number of antenna increases the losses in the phasing harness also unfortunately increase and the theoretical gains are not achieved. The stacking of dipoles is known as 'collinear'.

The parallel harness can be replaced with a series feed but again phase shifting coils are used to ensure that the aerials radiate in phase. However, with a series fed system more power is radiated from the lower antenna than the top and consequently the beam is tilted away from its centre. Altering the phasing allows the tilt to be controlled. By purposely tilting the beam downwards this effect can be used to advantage where close coverage within a mobile area is required or cell areas are small and spillage across a boundary has to be avoided.

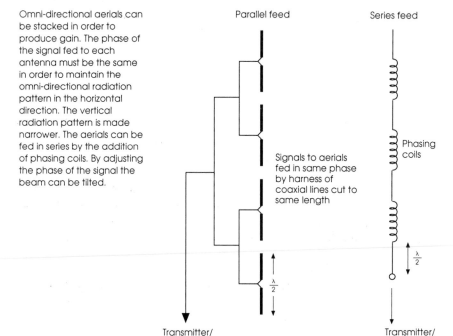

Omni-directional aerials can be stacked in order to produce gain. The phase of the signal fed to each antenna must be the same in order to maintain the omni-directional radiation pattern in the horizontal direction. The vertical radiation pattern is made narrower. The aerials can be fed in series by the addition of phasing coils. By adjusting the phase of the signal the beam can be tilted.

Parallel feed

Series feed

Signals to aerials fed in same phase by harness of coaxial lines cut to same length

Phasing coils

$\frac{\lambda}{2}$

$\frac{\lambda}{2}$

Transmitter/receiver

Transmitter/receiver

Figure 8.17   *Collinear arrays.*

Figure 8.18 *Collinear reflector. Angle either 60 or 120 degrees.*

In practical systems a four-stack dipole array can be used, although this can produce a ragged omni-directional pattern, or eight dipoles spaced 0.8 wavelengths
apart and installed in a four-H arrangement can produce an elliptical radiation pattern suitable for mobile coverage within a town.

Aerial gain can be obtained with a collinear array at the expense of an omni-
directional radiation pattern if a reflector is placed behind the array. The reflector is
constructed with vertical rods spaced at intervals of approximately one wavelength
of the operating frequency. As in all the VHF and UHF aerials dimensions are critical for optimum operation. In order to obtain an omni-directional pattern several
arrays must be spaced around a mast. The reflectors are shaped to provide a 120 or
60 degree corner reflector and, therefore, either three or six antennae must be
installed on a mast to obtain the 360 degree coverage (see Figure 8.19).

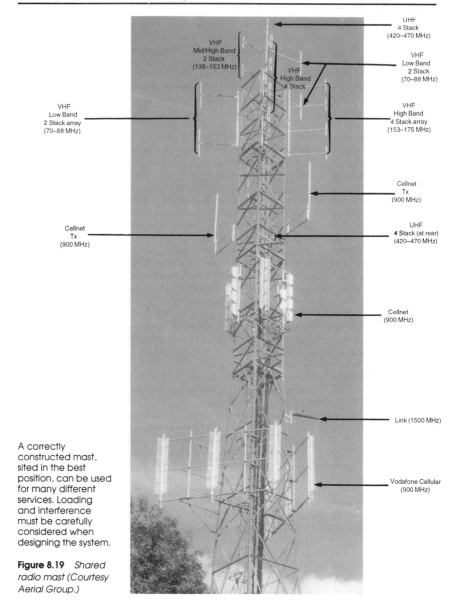

UHF
4 Stack
(420–470 MHz)

VHF
Mid/High Band
2 Stack
(138–153 MHz)

VHF
Low Band
2 Stack
(70–88 MHz)

VHF
High Band
4 Stack

VHF
Low Band
2 Stack array
(70–88 MHz)

VHF
High Band
4 Stack array
(153–175 MHz)

Cellnet
Tx
(900 MHz)

UHF
4 Stack (at rear)
(420–470 MHz)

Cellnet
Tx
(900 MHz)

Cellnet
(900 MHz)

Link (1500 MHz)

Vodafone Cellular
(900 MHz)

A correctly constructed mast, sited in the best position, can be used for many different services. Loading and interference must be carefully considered when designing the system.

Figure 8.19 *Shared radio mast (Courtesy Aerial Group.)*

Slot panel

This is another type of aerial used at base stations. It consists of three dipoles which are placed in front of a reflecting panel. The dipoles are fed in parallel at the top and bottom from a transmission line. Four such panels spaced 0.75 of the operating wavelength apart must be used around a mast to provide a 360 degree coverage. This type of aerial provides a relatively clean signal.

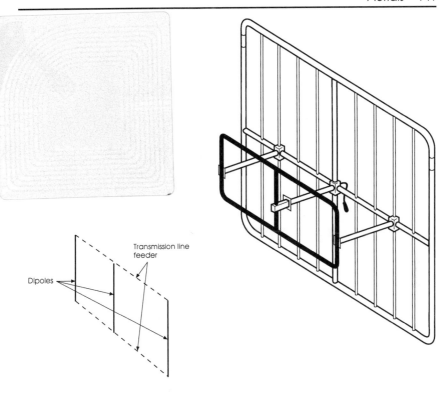

Dipoles

Transmission line feeder

Figure 8.20 *Slot panel.*

Shaping the radiation pattern

In order to increase the use of the radio spectrum the distances between areas reusing frequencies are being reduced. It is, therefore, important that directional aerials are able to separate the signals received from the intended direction and those received from the opposite direction. A directional aerial is, therefore, specified as having a 'front-to-back ratio' (in dB). The minimum requirement in Band III is 20 dB. Reflectors obviously improve the front-to-back ratio.

The radiation pattern of two omni-directional aerials in a bay formation is controlled by their spacing and the phasing of the signal supplied to the dipoles. By altering these parameters the design engineer can produce radiation patterns to suit different circumstances. By using a dipole spacing of 3/8 of a wavelength a long narrow figure of eight pattern can be obtained which is useful where a signal has to be confined. By using folded dipoles spaced a wavelength apart and adding an additional 1/4 of a wavelength to one of the aerial feeds, a cardioid pattern (heart shaped) is produced. In this latter arrangement, which can be stacked or bayed, the

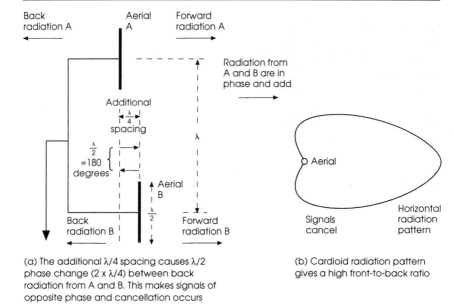

(a) The additional λ/4 spacing causes λ/2 phase change (2 x λ/4) between back radiation from A and B. This makes signals of opposite phase and cancellation occurs

(b) Cardioid radiation pattern gives a high front-to-back ratio

Aerial array used to provide a high front-to-back ratio.

Figure 8.21   *Quadrative phasing.*

signals in front are in phase and add while those in the rear are in anti-phase and cancel. This is known as quadrature phasing and can be used with both two and four stacked dipoles. A high front-to-back ratio of approximately 25 dB in Band III is achieved with this method.

Mobile antennae

The aerials previously described for base stations are stationary and few in number compared with the numbers of mobile antennae. Their positions and performance can be optimised and a type of aerial most suited for a particular situation can be chosen.

The mobile aerial is generally restricted to a single element which provides omni-directional coverage. Its position and method of installation on vehicles can, however, considerably affect the performance of the system.

Mobile radio uses vertical polarisation (Chapter 2). This allows the vertical aerial to be used and an omni-directional coverage achieved irrespective of the direction of the vehicle.

The vertical mobile aerials fitted to vehicles are of various lengths but each is directly related to the wavelength of the transmission frequency. All the aerials, however, require the body of the vehicle to provide the missing portion of the aerial in order to achieve acceptable transmission and reception. The vehicle must supply the ground plane for the aerial and as this requires a large metallic plate it is best achieved by fitting the aerial in the middle of the roof. If the antenna is mounted in any other position the body of the vehicle distorts the radiation pattern and the aerial ceases to become omni-directional.

If the vehicle does not have a metal body but is made of fibre glass or other material then it is necessary to provide a metal or foil base on which to mount the antenna. It is then necessary to provide an earth connection between the antenna and the ground plane.

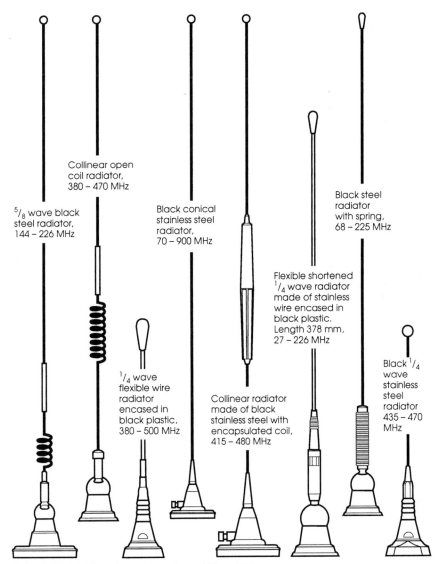

Figure 8.22   *Mobile antennae, (courtesy of Allgon Ltd).*

In addition to the position of the antenna affecting the performance it is essential that the antenna input impedance is correctly matched to the feeder cable in order to prevent standing waves and the loss of transmission power.

The simplest and most common antenna is the quarter wave antenna. From the centre of a vehicle's roof it radiates an omni-directional and even strength signal. Its impedance when connected to an even ground plane is approximately 40 ohms but this alters when the ground plane is uneven. This causes a standing wave when connected to a standard 50 ohm system. The antenna can be matched to the system by using the principle of a transmission line transformer as explained earlier in the chapter. In practice it requires the addition of a short length of coaxial cable which is calculated according to the frequencies and impedances of the system.

Larger antennae are able to intercept more signal. Even if only quarter wavelength antennae are compared, a VHF antenna is larger than a UHF antenna and performs better. However, a quarter wave antenna performs adequately at UK cel-

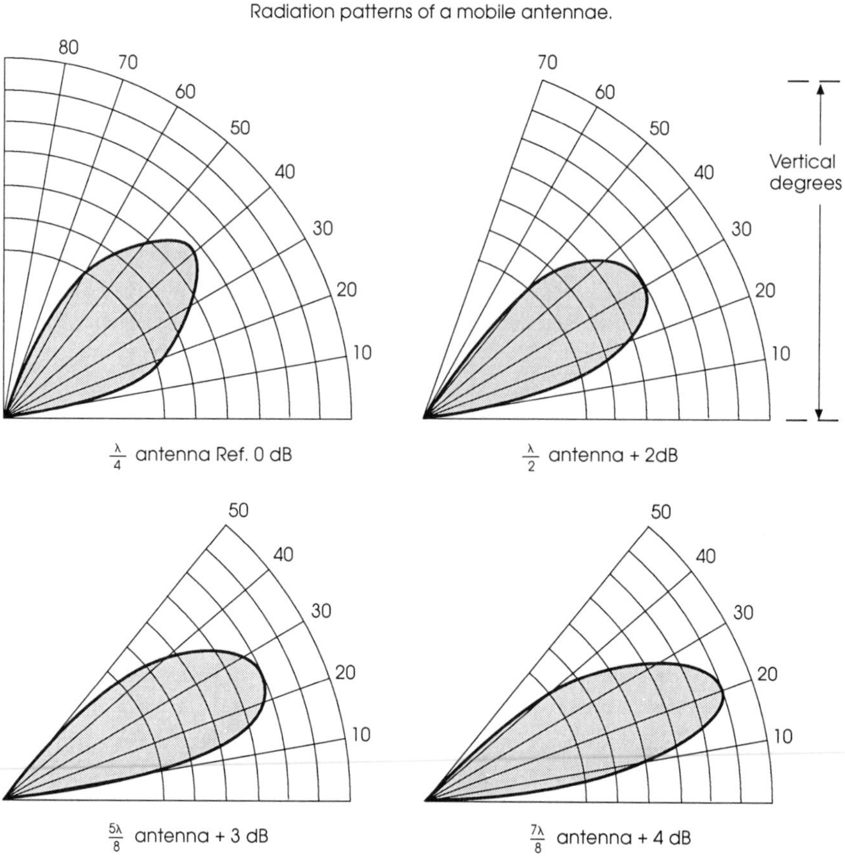

Radiation patterns of a mobile antennae.

$\frac{\lambda}{4}$ antenna Ref. 0 dB

$\frac{\lambda}{2}$ antenna + 2dB

Vertical degrees

$\frac{5\lambda}{8}$ antenna + 3 dB

$\frac{7\lambda}{8}$ antenna + 4 dB

Figure 8.23 *Radiation patterns from different sized antennae mounted centrally on car with average height of 1.5 m. Gain is achieved by making the beam narrower and more horizontal. These antennae are omni-directional and these patterns exist around the antenna.*

lular frequencies in most circumstances. In areas where base stations give poor coverage an aerial which provides some gain may be preferable. One which has been used for a long time is the 5/8 antenna. This means that the antenna length is increased to 5/8 of the wavelength of the operating frequency. The gain is achieved by the radiation being more concentrated in a horizontal direction and less skyward. This allows the signal to travel further and provides a theoretical gain of 3 dB when compared with the quarter wave antenna. The longer the antenna the lower is the angle of radiation. This is caused by the high current of the AC waveform occurring higher up the antenna, which alters the radiation angle. Other lengths of aerials used are half wave antenna (gain 2 dB) and 7/8 antenna (gain 4 dB).

In order to achieve a good standing wave ratio with the 5/8 antenna the aerial is fitted with a base loading coil. The impedance of the 5/8 line does not match the coaxial feeder but the addition of the windings artificially increases its length to a 3/4 wavelength which provides a suitable match with the feeder coaxial cable.

A disadvantage of the 5/8 antenna is a reduced bandwidth compared with a quarter wave antenna. The quarter wave has a bandwidth of approximately 10% of its operating frequency whereas this is reduced to approximately 8% for the 5/8 antenna. A larger physical size can also be a disadvantage in a fast moving car when the antenna can bend and distort the radiation pattern. Signal fluctuations can also occur with the larger antenna due to any swaying movement caused by the wind or movements of the vehicle.

A collinear antenna consisting of a 1/2 or 5/8 wavelength at the top and a 1/4 or 3/8 wavelength at the bottom which are joined by a coil is also popular. The aerial is end fed and produces a low horizontal radiation pattern and gain when compared with the 1/4 wavelength antenna. Its bandwidth is similar to a 5/8 antenna.

Glass mounted antenna

Glass mounted antennae have become popular in cars due to the fact that the car does not have to be drilled. They are normally fitted on the rear window and should be mounted centrally and as high as possible. They must be kept away from the window heating elements and at least a metre from any VHF car radio antenna. Before firmly mounting it should be established that they do not cause obstruction to the driver's vision or can be fouled by the rear wiper blade. They should be fixed at least 1 cm from the edge of the glass.

The heating elements, car radio antenna and the lead in glass can reduce the efficiency of the glass mounted antenna. Because also the antenna is not mounted in the best possible position (centre of the roof) the radiation pattern is affected and it is not completely omni-directional. Losses compared with an omni-directional antenna of the same type centrally mounted on the roof can vary between 1 and 2 dB in a forward and side direction and slightly more in the rear depending upon the shape of the car. However, the design of the antenna and the method of cabling can alter any measurements and a glass mounted antenna which is correctly designed and installed will provide better results than inferior equipment and a poor installation in the best position.

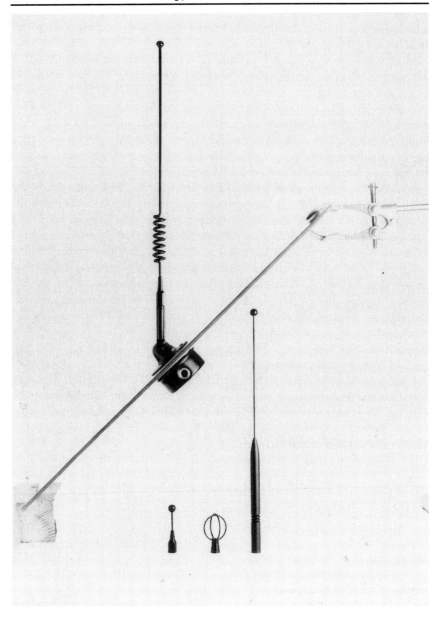

Collinear open coil radiator made of black chrome and plated steel. Gain is +4 dB relative to 1/4 wave radiator. Designed for mounting on glass. Operational frequency 872–960 MHz.

Figure 8.24   *Glass mounted mobile antenna. (Courtesy of Allgon Ltd.)*

The radiation pattern is almost the same as a λ/2 dipole. The input impedance is approximately 471 ohms

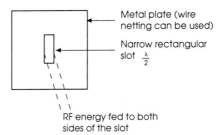

Metal plate (wire netting can be used)

Narrow rectangular slot $\frac{\lambda}{2}$

RF energy fed to both sides of the slot

Figure 8.25   *Slot antenna.*

Other types of antenna systems

In some instances it is not possible to use a conventional vertical whip antenna and in these circumstances other types or systems are used. Where height presents a problem, such as on double decked buses, loop or slot antennae are used and these are protected by radomes. Such systems are less than 63 mm high.

In order to provide communications by radio to trains three possibilities are in use. On straight tracks transmitters with directional antenna providing a narrow beam down the track can be used for phones. However, a large number of base transmitters and antennae are required if the track is not straight.

A leaky feeder antenna is used in Japan and runs at the side of the track. This is a coaxial feeder with slots in its outer casing which allow radiation of the signal. The radio signal is received by a horizontal mobile antenna which is mounted to the side of the train and runs alongside the feeder. With this method continuous communication is possible with the train.

It was explained in Chapter 2 that very long wavelength radio waves have the ability, due to their frequency, to penetrate areas irrespective of bends and obstacles. Use is made of this characteristic in underground systems to cover tunnels and passages.

Physical characteristics of aerials

Considerable thought must be given to the structures on which aerials are mounted at base stations. The thrust produced from an array of aerials when subjected to high winds is very high and calculations must be performed to ensure the mast can withstand the forces involved. Invariably the mast is situated on high ground and open to the full force of the weather. In addition the mast must be protected from lightning strikes with the lightning conductor being at least 2.5 m above the top of the mast and correctly earthed at the base.

Antenna components on masts can be protected against rain and snow with radomes. Materials for vehicle antennae must be selected to avoid corrosion as this affects both the mechanical and electrical properties of the antenna. Antennae should also be constructed to withstand shock, vibration and have some resistance to vandalism.

9 Radio paging systems

In 1956 the first radio paging system was installed in St Thomas Hospital in London. In 1972, the Post Office, who were then responsible for the regulations, licensed independent paging operators to set up their own networks. Since that time equipment, systems and facilities have grown considerably and Britain has remained the most advanced radio paging market in Europe. The market in Britain is expected to increase at a considerable rate each year in the future as, no doubt, it will throughout all the technically advanced countries.

There are obvious economic advantages to a radio paging system in comparison to a duplex communication system. The need to contact a person who can respond on a normal telephone may be sufficient for many businesses. The portable equipment has a relatively low cost and is small, lightweight and unobtrusive.

As only one channel is required the radio spectrum is used more efficiently and this can be further improved with digital and multiplexing techniques.

Pagers are offered in many forms. At the time of writing, message pagers are available which can receive messages with up to 200 characters and display 80 characters on a screen. Associated memories can also hold 40 such messages which can be protected or deleted as required. Messages are originated by a caller dictating a message to an operator at a systems bureau who then transmits the message to the pager. Advanced facilities allow a caller to be answered in the name of a subscribing customer or company. A message can also be simultaneously transmitted to several pagers in circumstances where a number of people have to receive the information. On line services also allow a subscriber of the individual services to receive financial news, details of traffic problems, etc.

Numeric pagers are available which display up to 20 numeric digits which can be telephone numbers, prices, product numbers, coded information or any other type of numeric data. They have a facility which allows a direct message input to the pager from a touch tone phone without the need to pass the message to an operator.

Tone pagers are the simplest and cheapest to hire and can be used to alert users to call predetermined telephone numbers or distinguish the urgency of a call when different

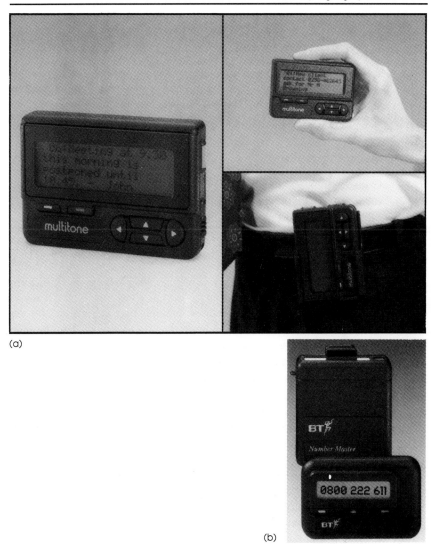

(a)

(b)

Figure 9.1 *(a) Alpha-numeric pagers (Courtesy Multitone Ltd).*
(b) Numeric pagers (Courtesy British Telecom).

audio sounds are received. Different types of pagers can alert users by tones, flashing light, vibration or even a silent mode when calls may be an intrusion during a meeting.

Voice messaging is also part of the new technologies and can instruct a caller to leave a message at a paging bureau for onward transmission or cell phone numbers which are not answered can be automatically transferred to a paging number. Voice transmission is also part of the on-site paging systems explained later.

Systems and equipment are constantly being improved and today's facilities may appear limited compared with those in the near future.

Types of paging systems

Two distinct types of systems are provided for paging purposes. These are:

1. Local area paging or on-site paging.
2. Wide area paging which uses a network of radio transmitters to provide paging within regional zones or nationally. This will be extended to Europe and eventually will become a global facility.

On-site radio paging

These systems are used within an organisation's premises and are designed to alert and inform people by using either singularly or in combination speech, data or bleeps. Four different types of on-site paging systems are in use. These include:

1. Induction loop paging
2. HF and VHF on-site paging
3. UHF on-site paging
4. Local communications

These systems are designed to provide radio communication only within the area controlled by the user of the system. This limits the area to a shop, factory, office, hospital, hotel or similar premises.

The frequencies used and the system characteristics are as follows (see Figure 9.2 for the actual frequency allocations).

1. Induction loop paging (frequencies used 16 to 150 kHz)
These systems are suitable for sites up to approximately 4 hectares (10 acres) with up to six buildings close together between which an induction loop can be economically run. Outgoing speech is permitted but return speech, if required, must operate in the 161 MHz band. The Radiocommunications Agency exempt low power induction loop systems from a licence requirement providing they are not two-way induction loop paging systems.

The induction loop used in a paging system is typically a leaky coaxial cable from which the radio energy radiates within the confined areas. Originally these systems used low frequency carrier signals and each receiver was tuned into its own carrier frequency. The capacity of the early systems was limited by the bandwidth of the system and the spacing between the channels. The capacity is increased when a single carrier is used which is FM modulated by various signals.

Although not a paging system, the induction loop is proving of considerable benefit to people with hearing difficulties. A simple wire loop around a room allows amplified signals from a unit connected to the TV or sound system to be received either by a special listening unit or the T position on the hearing aid. Many public buildings have been equipped with induction loops to assist those people with hearing difficulties to enjoy the theatre or concerts.

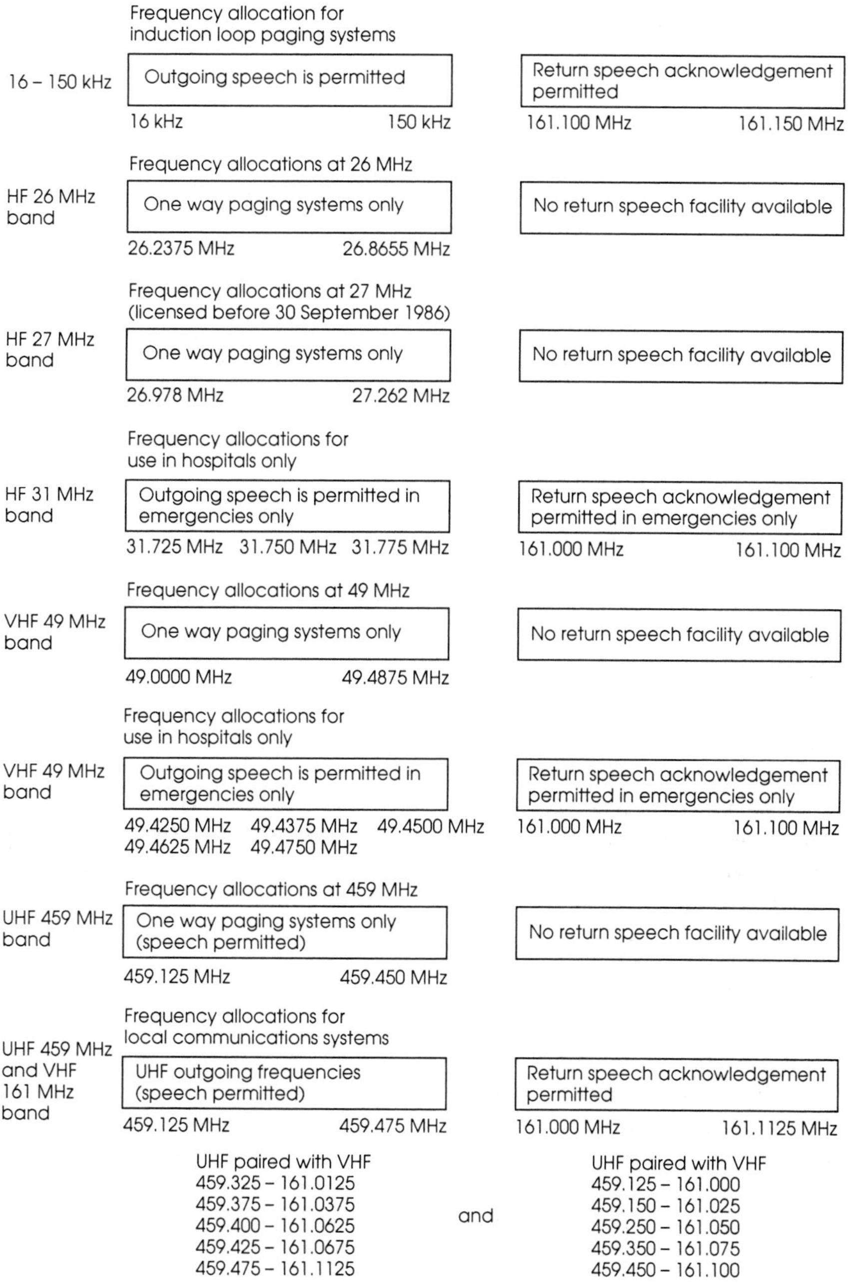

	Frequency allocation for induction loop paging systems	
16 – 150 kHz	Outgoing speech is permitted	Return speech acknowledgement permitted
	16 kHz 150 kHz	161.100 MHz 161.150 MHz

	Frequency allocations at 26 MHz	
HF 26 MHz band	One way paging systems only	No return speech facility available
	26.2375 MHz 26.8655 MHz	

	Frequency allocations at 27 MHz (licensed before 30 September 1986)	
HF 27 MHz band	One way paging systems only	No return speech facility available
	26.978 MHz 27.262 MHz	

	Frequency allocations for use in hospitals only	
HF 31 MHz band	Outgoing speech is permitted in emergencies only	Return speech acknowledgement permitted in emergencies only
	31.725 MHz 31.750 MHz 31.775 MHz	161.000 MHz 161.100 MHz

	Frequency allocations at 49 MHz	
VHF 49 MHz band	One way paging systems only	No return speech facility available
	49.0000 MHz 49.4875 MHz	

	Frequency allocations for use in hospitals only	
VHF 49 MHz band	Outgoing speech is permitted in emergencies only	Return speech acknowledgement permitted in emergencies only
	49.4250 MHz 49.4375 MHz 49.4500 MHz 49.4625 MHz 49.4750 MHz	161.000 MHz 161.100 MHz

	Frequency allocations at 459 MHz	
UHF 459 MHz band	One way paging systems only (speech permitted)	No return speech facility available
	459.125 MHz 459.450 MHz	

	Frequency allocations for local communications systems	
UHF 459 MHz and VHF 161 MHz band	UHF outgoing frequencies (speech permitted)	Return speech acknowledgement permitted
	459.125 MHz 459.475 MHz	161.000 MHz 161.1125 MHz

UHF paired with VHF		UHF paired with VHF
459.325 – 161.0125		459.125 – 161.000
459.375 – 161.0375		459.150 – 161.025
459.400 – 161.0625	and	459.250 – 161.050
459.425 – 161.0675		459.350 – 161.075
459.475 – 161.1125		459.450 – 161.100

(Prepared by Radio Communications Agency in conjunction with Radio Paging Association.)

Figure 9.2   *Frequency allocations for on-site paging services in the UK.*

2. HF systems (26 and 31 MHz) and VHF systems (49 MHz)

These frequencies provide good coverage within a site and are popular with system designers. The paging is provided by direct radiation. Since July 1985 all new licences are for non-speech paging only and must consist of data or bleeps. Part of the spectrum from both the 31 MHz and the 49 MHz bands has been set aside exclusively for hospitals in order to avoid interference from diathermy equipment or from other local radio systems. Outgoing speech can be used by hospitals on their allocated frequencies in the 31 and 49 MHz bands with return speech on 161 MHz providing it is for an emergency. Other special arrangements exist for two way HF paging systems existing before July 1985 or modified to add a return speech acknowledgement since that date. In 1997 new regulations will be enforced whereby such equipment must have been modified to conform to communication specification MPT 1314.

3. UHF systems (459 MHz)

In heavy reinforced concrete buildings and manufacturing plants, HF systems may be difficult to use and UHF may provide a better coverage. Data and coded bleeps are used and outgoing speech is also permitted but not return speech for acknowledgement.

4. Local communications (459 MHz for outgoing speech and paging, 161 MHz for speech acknowledgement)

Local communications is a relatively new radio system which was initiated in 1979 and first licensed in 1982. It is intended that it will eventually replace all existing on-site HF and UHF two way paging systems apart from those in hospitals and induction loop paging systems. It allows users to acknowledge a paging call by using speech in addition to outgoing speech and this feature provides two way speech facilities within the user's operating area for a radius of approximately 1 km.

Channel spacing

On-site paging systems are used in a confined area and are designed as low capacity systems ranging from typically 10 to 2000 radio pagers. The channel spacing in the VHF and UHF bands is 25 kHz although there are 12.5 kHz channels in the 49 MHz band. A narrower channel makes the system less sensitive and, therefore, receivers become more expensive.

In order to confine radiation the transmitter power is kept to a minimum and 25 mW ERP is typical. The aerial is installed to produce a conical beam which radiates downwards and a horizontally polarised crossed dipole can be used as illustrated in Figure 9.4. On-site paging may, however, involve more than one type of system and radiation, induction loop, intercoms and telephones may all be used to make a comprehensive and efficient hybrid system.

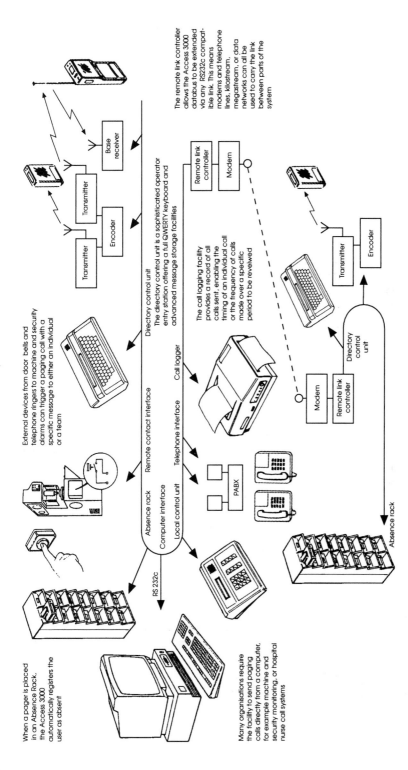

The remote link controller allows the Access 3000 databus to be extended via any RS232c compatible link. This means modems and telephone lines, kilostream, megastream, or data networks can all be used to carry the link between parts of the system

The directory control unit is a sophisticated operator entry station offering a full QWERTY keyboard and advanced message storage facilities

Directory control unit

External devices from door bells and telephone ringers to machine and security alarms can trigger a paging call with a specific message to either an individual or a team

The call logging facility provides a record of all calls sent, enabling the timing of an individual call or the frequency of calls made over a specific period to be reviewed

Base receiver

Transmitter

Transmitter

Encoder

Remote link controller

Modem

Transmitter

Encoder

Directory control unit

Modem

Remote link controller

Call logger

Absence rack

Remote contact interface

Computer interface

Telephone interface

Local control unit

RS 232c

PABX

Absence rack

When a pager is placed in an Absence Rack, the Access 3000 automatically registers the user as absent

Many organisations require the facility to send paging calls directly from a computer, for example machine and security monitoring, or hospital nurse call systems

Figure 9.3 A sophisticated system enabling a variety of inputs to the paging system, is provided by a local area network (LAN, Chapter 7). Remote areas can be connected by the use of lines and all the facilities extended. (System diagram courtesy of Multitone Ltd – Access 3000 System.)

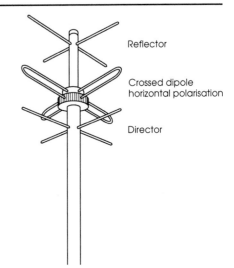

Reflector

Crossed dipole
horizontal polarisation

Director

Figure 9.4   *UHF paging antenna for on-site
systems. This type of antenna produces a
conical beam which radiates downwards.
This confines the radiated energy to the
system paging area.*

Wide area paging systems

A wide area paging system allows the transmission of one-way calls via a base station
to receivers within a defined area of the base station. The distance varies considerably between 5 and 20 km from the base station depending upon the location,
power and aerial system of the base station and the location of the receiver.

The signal is attenuated by buildings and the human body itself especially at high
frequencies. Tests have shown that frequencies between 80 MHz and 460 MHz are
suitable for radio paging in densely built areas. The attenuation of signal levels
caused by penetrating buildings ranges from 14 to 22 dB at 150 MHz, 18 dB at
250 MHz and 12–18 dB at 400 MHz.

Other factors which affect the ability to receive a signal include:

1. The level of environmental noise. Man-made noise in a city which can interfere
with the signal is inversely proportional to the frequency and, therefore, the higher
transmission frequencies are less affected.
2. The sensitivity of the receiver. This is directly related to circuit design and the
efficiency of the aerial.
3. The amount of paging traffic on the system.

Wide area systems can range from a small company system comprising a few pagers
operating in a limited area to national systems operating with thousands of pagers.

Frequencies used (Figure 7.2, Chapter 7)
Wide area paging systems operate in both the VHF and UHF bands at 138 and
153 MHz (VHF) and 454 MHz (UHF). Display calls are permitted on both bands.
However, the VHF band is limited to tone calls but either tone only or tone and
voice calls are permitted on UHF.

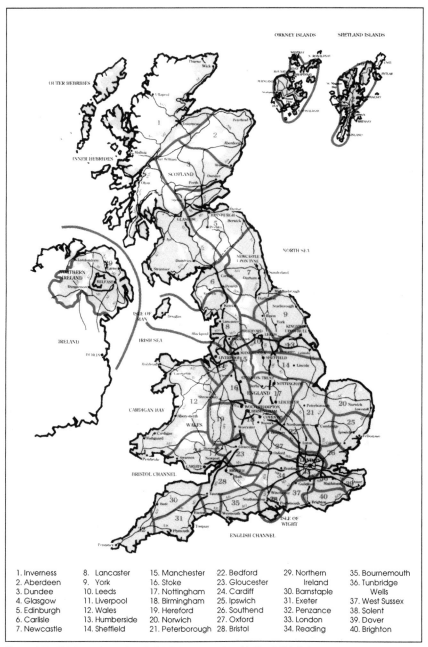

1. Inverness	8. Lancaster	15. Manchester	22. Bedford	29. Northern	35. Bournemouth
2. Aberdeen	9. York	16. Stoke	23. Gloucester	Ireland	36. Tunbridge
3. Dundee	10. Leeds	17. Nottingham	24. Cardiff	30. Barnstaple	Wells
4. Glasgow	11. Liverpool	18. Birmingham	25. Ipswich	31. Exeter	37. West Sussex
5. Edinburgh	12. Wales	19. Hereford	26. Southend	32. Penzance	38. Solent
6. Carlisle	13. Humberside	20. Norwich	27. Oxford	33. London	39. Dover
7. Newcastle	14. Sheffield	21. Peterborough	28. Bristol	34. Reading	40. Brighton

Figure 9.5   *Wide area paging (British Telecom network). The British Telecom paging network reaches approximately 97% of the population. The country is split into 40 zones and a customer is able to select the coverage most suitable for their personal requirements. The transmitters operate on four different frequencies at 153 MHz with channel spacing of 25 kHz. (Courtesy British Telecom.)*

Many local private wide area paging systems exist in addition to several systems which operate nationally for the general public. A system requires an exclusive frequency within its paging area and due to demand and a limited spectrum, the same frequencies may be allocated in adjacent system areas. In these circumstances it requires careful system engineering to avoid interference on the fringes of the areas.

In the 153 MHz band three manufacturers have been allocated preferred frequencies. These companies are Multitone, Motorola and Philips Telecommunications. Applicants must, therefore, make known to the Radiocommunications Agency the manufacturer of their equipment in order to be allocated the correct frequency.

Reception from two transmitters

The pagers, because of their compactness, contain a small built-in antenna. This is typically a small ferrite rod and coil which has a loss factor of approximately 16 dB relative to a half wave dipole. In order to compensate for the inefficient antenna and buildings and body losses, relatively high field strengths must be produced in order for the pager to function satisfactorily. This can be achieved by an additional transmitter simultaneously radiating the signal into the area. However, if both signals, which have the same carrier frequency and information, are received by the pager over different paths they must be timed in order to be received in phase, otherwise fading takes place.

Quasi synchronous operation

The method used to overcome this problem is known as quasi synchronous operation. As the transmitters are separated by different distances from the originating source it is necessary to add additional delay to the transmission path which feeds the nearest transmitter. When the two transmitters radiate, the signals are exactly in phase irrespective of the distance they have travelled to the transmitter.

When the pager receives both signals the ideal reception conditions exist when they are in phase. However, although the additional delays compensate for differences in land lines and radio links feeding the transmitter in addition to differences in equipment delays, they do not compensate for propagation differences due to the position of the pager in the area.

In a built-up area VHF and UHF signals are reflected from the buildings. The direct and reflected signals interact with each other to produce complex interference patterns and should a 180 degree phase difference between the signals occur, they can cancel each other. It is found, however, that by using the quasi synchronous systems the null pattern is different and total signal loss does not occur due to the interference patterns produced.

In Chapter 12 data systems are explained and it will be seen that frequency shift keying (FSK) is used for data transmission. A digital 1 or 0 is represented by deviation of the carrier frequency above or below the carrier frequency. The system, therefore, has two individual frequencies to represent 1 or 0.

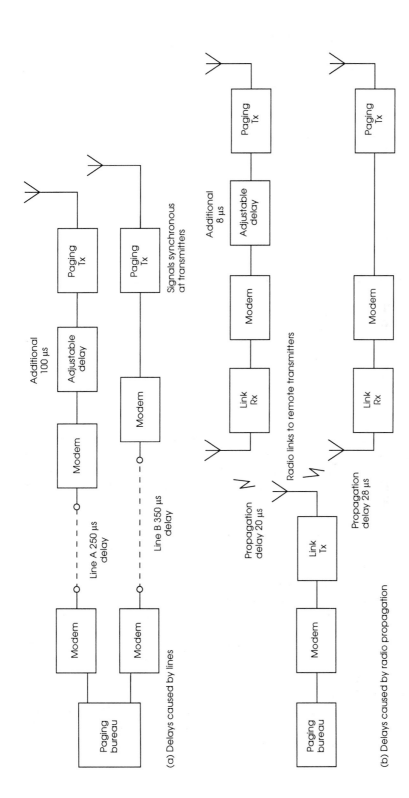

Figure 9.6 *Quasi synchronous operation.*

These data signals are able to be received by the pager which is more tolerant of distortion caused to them by reflection patterns. However, if speech is transmitted a band of frequencies between 300 and 3400 Hz needs to be received and the problem becomes more difficult.

A 1 kHz signal has a wavelength period of $1/f = 1$ ms. If a delayed and a direct signal have a 1 ms delay between them, the two 1 kHz signals remain in phase. However, a 500 Hz signal has a wavelength period of $1/f = 2$ ms. If the same delay is experienced the direct and delayed signals are 180 degrees out of phase and cancel. This produces selective fading (Chapter 2).

From experience it is found that the maximum permissible delay for speech is approximately 10% of the period for the highest transmitted frequency. For 3400 Hz the period is $1/f = 1/3400$ s. This equals approximately 300 microseconds and 10% is 30 microseconds.

For a binary signal there is a period known as the 'confusion period'. This is defined as the difference in time between the two transmitters changing from 0 to 1. The practical permissible delay time is given as 25% of any one-bit period. Both 512 and 1200 bits per second are used and consequently the maximum delays are:

1. $1/512 \times 25\%$ = 488 microseconds.
2. $1/1200 \times 25\%$= 208 microseconds.

In addition to any delays involved in links and equipment the position of the pager in relation to the two transmitters also introduces more delays and increases the confusion time. Radio waves, as shown in Chapter 2, travel at 3×10^8 metres per second which is the equivalent of 3.33 μs/km. However, because the modern pager uses frequency modulation, the capture effect occurs whereby the pager becomes locked onto the more powerful signal as the difference between the distances of the transmitters and the pager becomes greater. Once the capture effect occurs and the pager is operating from only the strongest signal, the delay effect between the transmitters is no longer a problem.

Digital code for wide area paging – POCSAG

In 1976 an international group of engineers met in London to produce a specification for wide area paging. As it was chaired by the Post Office the acronym POCSAG was given to the eventual specification – Post Office Code Standardisation Advisory Group. International recognition was given to the code in 1981 when it was accepted by the CCIR as the recommended Radio Paging Code No. 1 (RPC No. 1), (Rec. 584).

Features of the code
1. Code can support alert only pagers, numeric pagers and alpha-numeric pagers.
2. Two million addresses available which is increased to 8 million when function bits are used to select the address from the four assigned to a pager.
3. Error detection codes used.

4. FSK modulation used in non–return to zero manner; +/–4.5 kHz deviation of the carrier with a 25 kHz channel (+dev. 0,–dev.1). FSK maximises the tolerance for time misalignment of simultaneous transmitters.

5. Air time can be maximised by a mixture of alert only and longer messages.

6. Battery saving with an idle mode.

7. High successful call rate.

POCSAG radio paging code

As shown in Figure 9.7, a transmission commences with a preamble. This is a series of reversals (101010 etc.) of at least 576 bits. This is followed by batches of complete codewords. Each batch consists of a synchronisation codeword followed by eight frames each of which contains two codewords (total of 17 codewords in a batch).

The eight frames are numbered from 0 to 7. The total number of pagers on the system is divided into the eight groups. Each pager has a 21-bit identity and the actual group it is allocated to depends upon the three least significant bits (e.g. 000 frame 0, 111 frame 7). The pager only examines the address codewords in its allocated frame and, therefore, its address codewords must only be transmitted in the allocated frame. This allows battery saving as the receiver is only turned on during the sync codeword and its own particular frame (only 3/17 of time required for constant reception).

A message can consist of any number of codewords spread over several batches but:

1. They must not displace sync codewords.

2. They must follow directly the address codeword.

3. The message is terminated by the next address or idle codeword. One or the other appears before the next message. Whenever there is no meaningful codeword in a batch an idle code is transmitted (Figure 9.10).

When transmitting at a rate of 512 bits/s a batch takes only 1.0625 seconds to transmit. The transmission can then be stopped. This allows the engineering of Time Slotted Multi-transmitter systems. These can be arranged as low cost sequential systems or the transmitters can be zoned and the frequency reused. In non-time-slotted systems it is possible to engineer priority calling with delays of approximately 1 second.

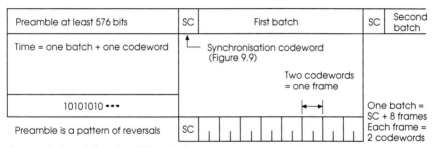

Figure 9.7 *Signals format, POCSAG code.*

Codewords (Figure 9.8)

Message and address codewords are arranged differently and are identified by a flag bit:

Bit 1 = 0 identifies an address
Bit 1 = 1 identifies a message

Each codeword consists of 32 bits and the most significant bits are transmitted first.

Address code

Each pager has a 21 bit code and has been allocated a particular frame depending upon the three least significant bits. The 18 most significant bits of the pager's identity are located as bits 2–19. Four addresses can be assigned to the pager and bits 20–21 select the required address. The total number of addresses now becomes over 8 million (2^{23}). Parity checks are provided by bits 22–31 and even parity is provided by bit 32.

Message Code

The message code is selected by a binary 1 at bit 1. Messages immediately follow an address code. The message codewords continue until a new address or idle codeword appears. Message codewords can continue into the next batch but the format of 16 codewords preceded by a synchronisation codeword is maintained. At least one address or idle codeword is displaced by the message and these are delayed and transmitted at the next available frame. The message codeword consists of 20 bits (2–21) followed by parity check bits (22–32). The maximum length of a message is governed by the storage capacity of the pager. The minimum storage has been standardised as 40 characters for alpha-numeric pagers and 20 characters for numeric pagers. Message formats for both numeric and alpha-numeric pagers have been standardised. The alpha-numeric format obviously needs a greater number of characters and 7 bits are used; 4 bits are used for the numeric format and this saves air time when transmitting.

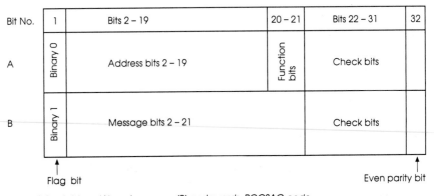

Bit No.	1	Bits 2 – 19	20 – 21	Bits 22 – 31	32
A	Binary 0	Address bits 2 – 19	Function bits	Check bits	
B	Binary 1	Message bits 2 – 21		Check bits	

↑ Flag bit   Even parity bit ↑

Figure 9.8 *Address (A) and message (B) codewords, POCSAG code.*

Bit No.	1	2	3	4	5	6	7	8	9	10	11	12	13	14	15	16
Bit	0	1	1	1	1	1	1	0	0	1	1	0	1	0	0	1	0

Bit No.	17	18	19	20	21	22	23	24	25	26	27	28	29	30	31	32
Bit	0	0	0	1	0	1	0	1	1	1	0	1	1	0	0	0

Figure 9.9 Synchronisation codeword.

Idle codeword (Figure 9.10)

The idle code is also a valid address codeword and consequently the number must not be issued to a pager. Its value is high and unlikely to be used as an address. Its equivalent numbers are 2007664–2007671. (Bits 2–21, see 'Paging code,' page 159.) A shift register could generate both the sunchronisation and idle codeword.

Bit No.	1	2	3	4	5	6	7	8	9	10	11	12	13	14	15	16
Bit	0	1	1	1	1	0	1	0	1	0	0	0	1	0	0	1

Bit No.	17	18	19	20	21	22	23	24	25	26	27	28	29	30	31	32
Bit	1	1	0	0	0	0	0	1	1	0	0	1	0	1	1	1

Figure 9.10 Idle codeword.

Codeword generation (known as 31:21 BCH + parity)

Each codeword consists of 31 bits of which 21 are the information bits. To the 31 bits is added one additional bit to provide an even parity check of the whole codeword. The bits which are not information bits are generated to provide checks for errors and their generation depends upon the information bits. By the use of Hamming codes (beyond the level of this book – refer to a specialist data coding book) comparisons can be made on reception and data errors of various lengths can be detected and possibly corrected.

System engineering for a digital radio paging system

There is a security advantage to transmitting digital messages to pagers as it is not possible to interpret the message without a sophisticated decoder. A listener to the radio channel will only hear bursts of coded information which will be unintelligible.

The POCSAG code, as previously described, is accepted as the transmission coding standard by most manufacturers and this allows a variety of pagers to be used when designing a POCSAG paging network.

The number of transmitters required to cover an area depends upon the topography of the area, the aerial height and the type and density of the buildings. A spacing of 5–8 km may be necessary in a densely built city centre while transmitters spaced 20–25 km apart may be satisfactory in a rural area. Transmitters are operated in quasi synchronous mode, as previously explained, in order to minimise distortion when a pager receives signals from more than one transmitter.

VHF high band is used by the majority of nation-wide systems as the coverage that it provides within cities and rural areas is a satisfactory compromise. A single 25 Hz channel is used in the 138–174 MHz band. The UHF band 420–470 MHz is also used but generally a UHF transmitter does not have the range of a VHF transmitter. UHF does, however, have the advantage that in a dense city centre the radio waves will penetrate the buildings better than VHF. All radio paging channels are dedicated channels and cannot be used for other purposes.

The transmitters are frequency modulated using FSK (Chapter 12) for the digital encoding. Data rates of 512 baud and 1200 baud are used and the higher data rate allows more pagers to be used on each radio channel. The data format is POCSAG.

Figure 9.11   *A paging bureau receives messages from callers for transmission to the paging subscriber. A variety of facilities are offered. The caller wishing to contact a pager only needs to dial the pager's exclusive number. The bureau answers in the name of the customer or company. One number can be used to contact a group of pagers and the message is transmitted simultaneously to all the necessary pagers. Messages can be stored for collection.*
When the pager number is called, the relevant information about the customer appears on the operator's VDU. The operator types the message for onward transmission to the pager. (Courtesy of Air Call Communications.)

Transmitters in the UK and in most European countries are limited to 100 watts effective radiated power (ERP) from the antenna for the transmission of digital radio paging signals.

System infrastructure

The design of the system must allow a message to be accepted, encoded into a digital format, stored and transmitted in a POCSAG paging format. The equipment required to perform these functions consists of:

1. Message entry equipment
2. Central computer
3. Transmission network equipment

Message entry equipment

Messages can be entered into the system by several methods. These include:

1. Operators at a bureau receive messages by telephone and retransmit them by keyboards.
2. Operators at remote terminals can enter messages which are transmitted via leased or PSTN lines.
3. Automatic paging can be initiated without bureau operators by an automatic direct dial facility. This requires a digital paging switch and offers the facility of operating tone, numeric and stored and pre-defined alpha-numeric messages by dialling numbers into the system via the PSTN.

Central computer

The central computer controls the system and becomes a message control unit. The messages are passed into it from the various interface units and are formatted and addressed. The message control unit incorporates a hard disc on which is recorded all

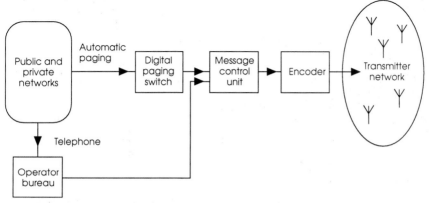

Figure 9.12 *Outline of a paging system. (Courtesy Paging Systems Ltd.)*

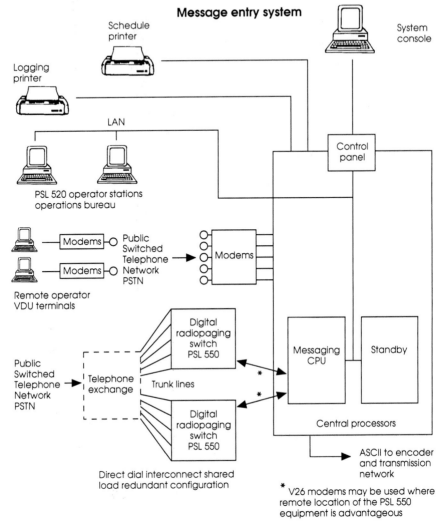

Figure 9.13 *Typical message entry system and transmission network for a paging system. (Courtesy Paging Systems Ltd.)*

the transmitted messages. These are kept for a number of days or stored more permanently onto tape. The disc also contains all the necessary software for the operation of the system. A back-up MCU may be provided due to the importance of the unit for the operation of the system. In complex systems additional remote MCUs may be provided which are connected together by a LAN (Chapter 7).

Transmission network equipment
Equipment is required for:

1. Message encoding and control
2. Transmitter control
3. Radio transmission synchronisation
4. Network management (optional)

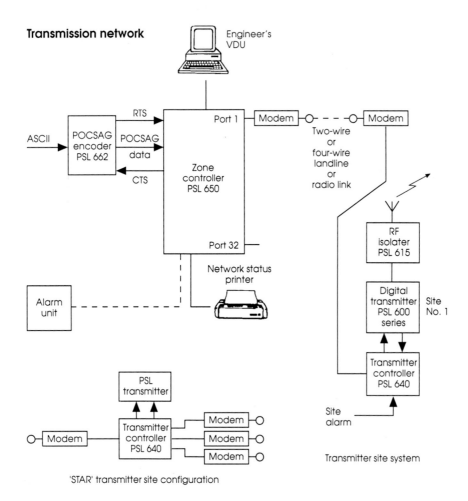

Figure 9.13 *(Continued)*

The POCSAG encoder receives the messages from the MCU in ASCII format. Input control commands determine the priority, urgent messages and batched and continuous transmission modes. Message storage buffers are organised as priority, urgent and normal groups and there is one group for each of the eight frames within the POCSAG code addressing structure. There are 64 kbytes of message storage which can be expanded to 1 Mbyte. The storage buffers ensure a high traffic rate can be obtained without the loss of messages. They also provide storage if interruptions occur on the system due to engineering work.

Paging messages of 1200 and 512 baud are automatically mixed by the encoder and it transmits the messages in POCSAG code to the transmitters once they are

Introduction

The PSL 600 transmitter is purpose designed for wide area digital radio paging networks. It is ideal for quasi synchronous (simulcast) transmission

An ultra high stability exciter employs synthesised frequency control whereby pre-set frequencies stored in PROM are digitally derived from a single 5 MHz reference crystal.

The compact design features simple maintenance with minimal adjustments and the modular synthesiser unit allows rapid module replacement.

An automatic RF power reduction system operates in event of over-temperature conditions.

RF output power is continually adjustable over the range 30 to 150 watts with 100 watts available for continuous duty and up to 150 watts available for intermittent duty operation.

A front panel digital meter and LED indicators provide local monitoring, together with an RF 'sample' output for maintenance purposes.

Comprehensive alarm and control interfacing facilities permit remote control and supervisory monitoring of the transmitter (and site) status conditions with both digital and analogue reporting.

Figure 9.14 *FSK digital radio paging transmitter type PSL 600. (Courtesy Paging Systems Ltd.)*

primed and ready to transmit. The command to transmit comes from the zone controller. Network management, if provided, is also a function of the zone controller which operates in conjunction with the transmitter controllers. Transmitter controllers are situated at every transmitter site and by continuous communication with a zone controller up to 99 transmitters can be controlled and remotely supervised. The management system also ensures that the correct compensation for propagation delay at each transmitter is inserted. Correction can be remotely controlled by commands from the zone controller or made on site.

The POCSAG idle codewords and the preamble which is at the start of each batched output are automatically generated and inserted by the encoder.

The transmitter controller function is the supervisory monitoring, control and data delay compensation of the transmitter and it receives its operating instructions from the zone controller. If the transmitter controller is not fitted the delays must be set manually. When multiple zones are required, each zone is fed from its own POCSAG encoder. Zone controllers and transmitters then function as independent but fully synchronised sub-systems. The digital FSK transmitters are crystal

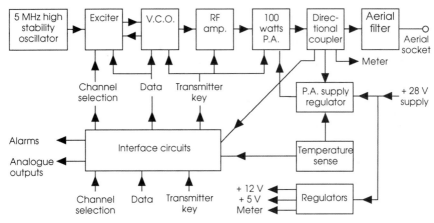

V.C.O. = voltage controlled amplifier
P.A.   = power amplifier

Features

- Ultra high stability, fully synthesised exciter

- Compact size, 132 (3u) H. 483 W. 370D

- 100 watts continuous, 150 watts intermittent

- Eight channel operation, 1 MHz bandspread

- Digitally controlled deviation 4.54 kHz or 5.00 kHz fixed

- Type approved to the UK specification MPT 1325

- Suitable for 512 and 1200 baud POCSAG

- Antenna output, short circuit and open circuit protection

- Over-temperature protection – with automatic RF output power reduction

- Front panel RF sample output

- Front panel digital meter

- DC powered 28 V at 12 amps (nominal)

- AC powered 110 V or 240 V via associated (2u) power supply unit PSL 610

- Alarm and control interface – compatible with PSL network management system

Figure 9.15   *FSK digital paging transmitter block diagram. (Courtesy Paging Systems Ltd.)*

controlled for high stability and this ensures optimum quasi synchronous operation. A continuous 100 watts ERP is radiated from the aerial via an isolator. This is inserted to reduce the level of intermodulation distortion produced by the transmitter when operating on communal sites where other transmitters are operating in the same frequency band.

10   Cellular radio systems

Several countries have produced analogue cellular radio systems and these have been mainly based on two systems: the American AMPS system (Advanced Mobile Phone System) and the Scandinavian NMT system (Nordic Mobile Telephone). It was in these countries that the initial developments occurred. The UK TACS system (Total Access Communication System), operated by Vodafone and Cellnet, is based on the AMPS system. The systems operated in Europe in January 1993 are shown in Table 10.1. Unfortunately, in these first generation systems the specifications are different and the systems are incompatible. It is intended that a new generation of digital systems will be compatible throughout Europe. However, the design of the analogue systems provides common features although the technical details may be different. These include:

1. A cellular structure for communication between the base station and mobile.
2. Frequency reuse amongst the cells.
3. Handover between cells as a mobile passes from one to the other.
4. A full duplex communication system.
5. Automatic direct dialling in both directions.
6. Automatic roaming whereby the registration of the phone on the system allows the user to roam nationally and eventually internationally.
7. Dedicated control channels used for setting up the call in most systems.
8. A continuous tone used to supervise the voice channel between the base station and the mobile in most systems.
9. As the number of users increases within a cell area, the cell can be split into smaller areas and the frequencies reused to allow for growth.

The services offered by a cellular network include:

1. Voice and data communication between mobiles and integration with the PSTN (Public Service Telephone Network).
2. The use of fixed car phones which provide maximum power, small hand portables for use in areas with strong cellular signals, transportable equipment which is

Table 10.1   *European analogue cellular radio systems (January 1993)*

System TACS-900 and ETACS Channel spacing 25 kHz, frequency band 900 MHz			
Country	Austria	Commenced	July 1990
	Ireland		December 1985
	Italy		April 1990
	Malta		July 1990
	Spain		April 1990
	United Kingdom		January 1985

System NMT-450 Channel spacing 25 kHz, frequency band 450 MHz			
Country	Andorra	Commenced	July 1990
	Austria		November 1984
	Belgium		April 1987
	Denmark		January 1982
	Faroe Islands		January 1989
	Finland		March 1982
	France		August 1989
	Iceland		July 1986
	Luxembourg		June 1985
	Netherlands		January 1985
	Norway		November 1981
	Spain		June 1982
	Sweden		October 1981

System NMT-900 Channel spacing 12.5 kHz, frequency band 900 MHz			
Country	Cyprus	Commenced	December 1988
	Denmark		December 1986
	Faroe Islands		June 1992
	Finland		December 1986
	Netherlands		January 1989
	Norway		December 1986
	Sweden		December 1986
	Switzerland		September 1987

System C-450 Channel spacing 20 kHz, frequency band 450 MHz			
Country	Germany	Commenced	September 1985
	Portugal		January 1989

System RC 2000 Channel spacing 12.5 kHz, frequency band 200 MHz			
Country	France	Commenced	November 1985

System RTMS, frequency band 150 MHz			
Country	Italy	Commenced	September 1985

System Comvick – commenced in Sweden in August 1981 but sales no longer made because of lack of radio spectrum. Only a small number of subscribers estimated at approximately 21 000

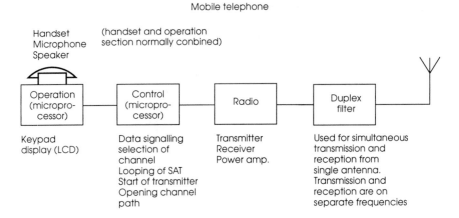

Figure 10.1   *It is the microprocessor which has made the mobile phone and the operational system possible. A mobile must tune to the best control channel. The mobile station logic unit automatically inserts the first control channel number into the frequency generator. The receiver senses if the quality of the channel is satisfactory and, if not, it will continue to scan the 21 control channels until a good signal is found. If a satisfactory channel is not found the mobile is not in a service area. Tuning from one frequency to the next takes approximately 20 ms. An additional microphone and speaker can be added for use in a car where hands-free operation is required.*

robust and more powerful and can be used anywhere within the cellular network, and the transmobile which is designed for use in cars or as a handportable.

Equipment and services are available to:

1. Forward messages which were left when the mobile user was unavailable.
2. Hands-free car phones.
3. Storage and automatic dialling of telephone numbers.
4. Voice operated dialling either by speaking numbers or by a previously stored company name.
5. Itemised billing for an individual or for a group of company cellular phones. Charges can also be displayed after a call.

Cellular phones can also be operated as coin-box mobiles in trains, coaches and taxis.

Typical system chart for a pocket carphone

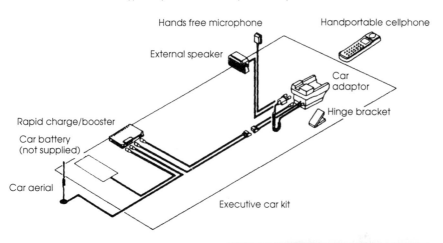

Hands free microphone

Handportable cellphone

External speaker

Car adaptor

Hinge bracket

Rapid charge/booster

Car battery (not supplied)

Car aerial

Executive car kit

Figure 10.2 *The mobile phone shown can be used as a handportable or installed in a car using the executive car kit. (Courtesy Cellnet.)*

Figure 10.3 *Transportable mobile phone. (Courtesy Cellnet.)*

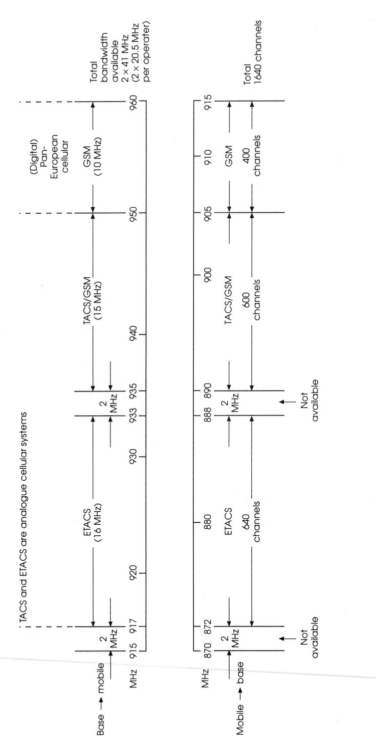

Figure 10.4 *UK allocation of frequency spectrum to TACS/ETACS/GSM. Operators are licensed to transmit GSM in the TACS bandwidth but not in ETACS.*

The available spectrum and channels

The problem with radio communication is that the radio spectrum available for any particular purpose is finite and cannot be automatically increased as demand increases with the growth of a system. In the UK the frequencies allocated are in the UHF band between 890 MHz to 905 MHz and 935 MHz to 950 MHz. These are known as the TACS network (Total Access Communications System) and are divided between Cellnet and Vodafone network operators.

The original number of channels quickly became congested in areas of high demand and the government made additional frequencies available in bands between 872 MHz to 888 MHz and 917 MHz to 933 MHz. These originally belonged to the Ministry of Defence. The additional channels are known as ETACS. Initially special phones were required to work in the ETACS bands but later generations of equipment are able to work in both TACS and ETACS wavebands.

Each channel has a bandwidth of 25 kHz with a 25 kHz spacing between channel frequencies. As each conversation is duplex two channels are required for each conversation and send and receive between the base station and the mobile are operated on different frequencies. To maximise the use of the spectrum and the number of available channels requires the reuse of the frequencies. However, a system must be devised that allows complete coverage within a defined area and for no interference to take place between transmitters operating on the same channel frequencies. This requires separation and control of both the power and direction of the radiated signal.

The initial 1000 channels were divided with 300 channels each for Vodafone and Cellnet and 400 reserved for the digital GSM system (see later in the chapter). The additional ETACS gave a further 320 channels to each operator from the 640 available channels.

The hexagon shaped cell

In order to provide a good grade of service in a cellular system the maximum number of channels must be available where the calling rates are high. Where calling

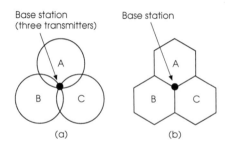

Figure 10.5 *(a) Three transmitters radiating into three cells A, B, C. (b) Hexagon representation.*

rates are low a fewer number of channels are required. This means that the size of the areas in which a specific number of channels are used is variable. A high calling rate area, such as a city centre, is divided into small cells in which a number of different frequencies are used for the channels. Adjacent to these are cells with transmitters operating on different frequencies. The transmitters in adjacent cells, therefore, do not interfere with each other. At a suitable distance, which is sufficient to avoid interference, the frequencies can be reused in other cells.

The size of cell when a cellular system is first installed is usually large and may be typically between 20 and 40 km in radius. It is determined by the signal-to-interference (S/I) levels which will be affected by signal strength, power, topography and radio interference. As the use of the system increases, the grade of service is reduced. Cell size is, therefore, reduced and the number of channels in a particular area is increased. Cell sizes can be as low as 0.5 km in radius where there are many users.

The shape of a cell must be such that total coverage is obtained by the cells interlocking. The ideal shape that allows this to occur is a hexagon and this shape is assumed for planning and the representation of cells on paper in order to simplify the situation when covering a specific area. In reality the shape is determined by the radiation pattern of the transmitter's antenna and may be an omni-directional antenna in rural areas. The complete coverage is known as cellular and provides the term cellular radio.

Planning cellular radio cell size and frequencies

When planning a system the aim must be to achieve the maximum use of the available radio spectrum. In addition there must be low interference, good quality speech and an acceptable grade of service.

In order to maximise the use of the system each cell should contain a large number of customers and, preferably, the traffic needs to be as even as possible throughout the day although this situation rarely arises (see Chapter 7). Channel frequencies are allocated according to the traffic demand and the number of channels in the different cells can vary. Maximum use of the spectrum is obtained by reusing the channel frequencies in other cells but they must be spaced at a sufficient distance from each other that they do not cause co-channel interference.

The 300 channels allocated to each of the two cellular systems in the UK TACS network allow 277 channels to be used for speech, 21 channels to be used for control signalling and 2 are used for guard channels between the Cellnet and Vodafone bands of frequencies. Although control channels cannot be used for speech, speech channels can be used for control if required.

The 320 channels for each operator in ETACS can all be used for speech as the same control channels are used for both TACS and ETACS.

The available channel frequencies are used in a varying number of cells known as a cluster. Only cells which carry a large amount of traffic are allocated the maximum number of channels and some rural cells may have very few channels. It all depends upon demand. Frequencies are reused in other clusters but are not repeated in a cluster. As the channels are full duplex each conversation requires a pair of frequencies. The forward and reverse directions from the base station to the mobile fall into two bands and the two frequencies used for a pair are separated by 45 MHz.

Clusters and cell size

Three factors affect the number of channels that can be used in a particular area:

1. The frequency spectrum that is available.
2. The smallness of the cells that can be achieved by careful engineering of transmitter power and control of radiation pattern.
3. The reduction in the quality of the link that can be tolerated due to co-channel interference.

Methods of trying to increase the number of channels by reducing the bandwidth of a channel (e.g from 25 kHz to 12.5 kHz) are counter productive as the protection against co-channel interference provided by the FM analogue system decreases and the spacing required before the channel frequency can be reused is increased. This

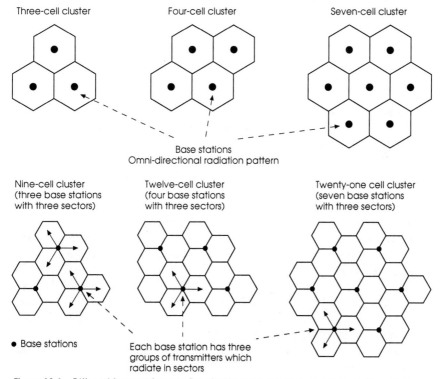

Figure 10.6 *Different frequencies are allocated to each cell and are not reused in a cluster. Cells in adjacent clusters which use the same frequencies must be separated and radiated-power adjusted to avoid co-channel interference. Channel frequencies within a cluster are allocated such that a cell does not contain adjacent channels.*

increases cluster size and reduces the channel capacity in an area. The distance required before the channels can be reused is calculated both theoretically and empirically and is determined by signal strengths in a particular environment.

The normal maximum number of channels operating in a cell is limited to 90 (maximum possible is 120) and this occurs in places where traffic is high such as the City of London. Cells in other situations carry much fewer channels. The capacity of a system in an area is determined by the number of channels in a cell and the cell size and repeat pattern. Various patterns are used and it is the carrier/interference ratio (C/I) over the area that determines the pattern. The recommended C/I ratio

Figure 10.7   *Cellnet's masts in Oxford Street, London. A mobile radio omni-directional and a sectored antenna are shown. On the roof are satellite receivers. In the background is the BT London Telecom Tower with microwave dishes for the BT national networks and these carry data, telephony and television. (Courtesy Cellnet.)*

Figure 10.8 *Cellnet mast and antennae in a rural area of Wales. (Courtesy Cellnet.)*

for TACS is 18 dB (11 dB for GSM – see later) although the system is capable of performing satisfactorily with a lower ratio of 12 dB. The operators have taken advantage of this by placing sites closer together or using a four-site repeat pattern which allows a given capacity to be provided by fewer sites but with a slightly lower C/I ratio.

In the typical patterns shown in Figure 10.6 the base station site is either central to the cell and radiates an omni-directional pattern or the site is sectored for 120 degree radiation patterns from three separate transmitters radiating into the separate cells.

Figure 10.9   *The overlaid cell uses a transmitter on the same site as the standard cell. The additional voice channels are used for small areas containing a high traffic demand. The same control channel is used for both the underlaid and overlaid cell.*

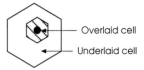

A typical 21-cell cluster, used in the UK TACS network, can be serviced from seven base station sites each with three sector cells. This arrangement provides the most cost effective system and the best capacity. In Central London a four-cell pattern is provided with 90 channels per cell on the Cellnet system and cell spacing is as close as 1 km.

Overlaid cells

In areas of very high traffic demand additional channel capacity may be required within a cell and in these circumstances an overlaid cell is provided at the point where demand is greatest. The overlaid cell uses frequency groups which have previously been allocated to other distant macrocells. A separate control channel is not provided for the overlaid cell and this facility is shared with the underlaid cell. The provision of an overlaid cell adds to the complexity of the system and requires base station software to allocate channels to mobiles and provide handover between the overlaid cell and the underlaid cell depending upon their signal levels. The transmitters for the overlaid cells radiate at a lower power than the underlaid cells and only mobiles extremely close to the base station are allocated channels on the overlaid cell. These mobiles are recognised by a high signal level.

Microcells

Microcells offer an engineering alternative in providing a large number of channels in areas where traffic is particularly dense. In these systems the antenna is below the average building height and, therefore, the signal radiation is restricted to a small area but one which contains many subscribers. Frequencies used by the macrocell are lost within the general clutter and can, therefore, be reused without degrading the quality of the service.

Aerial height

In a mature cell system served by several aerials a problem can arise when aerial heights are considerably different. A high aerial can capture all the traffic and become congested while other lower sites have spare capacity. In these circumstances the traffic has to be redirected to the surrounding sites. The situation can also be eased by restricting the coverage of the antenna by pointing it downwards to reduce the distance travelled by the radio waves.

Supervisory audio tones

The major problem with frequency reuse is the possibility of co-channel interference. Precautions include spacing sufficiently apart the cells which use the same frequencies, adjusting the power of the transmitters and controlling the radiation pattern of the base station aerials. An additional control is the use of supervisory audio tones (SATs). All the voice channels in a particular cluster are allocated one of three audio tones. These are 5970 Hz, 6000 Hz or 6030 Hz. Adjoining clusters have different tones. The allocated voice channel and the SAT frequency are determined by the mobile switching centre. The mobile receives the tone from the base station and this is looped back to the station for the duration of the call. If at any time during the call the SAT received from the mobile by the base station is different to the expected tone, due to the co-channel signal being received, then a call release or handoff is immediately requested.

Supervision of the Call

The progress of a call is continually monitored for both signal-to-noise ratio and the reception of the correct SAT tone. The tone is also used to provide a handoff request or release if the noise on the link reaches certain predetermined levels. The two levels are known as:

1. SNH. The is a level of signal-to-noise ratio which produces a handoff request. This request may or may not be possible to accomplish.
2. SNR. This is the level of signal-to-noise ratio which produces a call release request. Between SNH and SNR levels the quality of speech continually deteriorates and an SNR level prompts a call release even if a handover is not possible.

Signal strength monitoring

In addition to monitoring the SAT tones, the signal strength received from the mobile by the base station is compared with threshold levels predetermined by the mobile switching centre (MSC). These levels control the power from the mobile. Only the minimum power, sufficient for good reception, is required from the mobile. This is determined by the distance and the environment between the mobile and the base station. Excess power may cause interference, especially to other co-channel users. There are four predetermined levels:

1. SSD. This is a request for a Signal Strength Decrease.
2. SSI. This is a request for a Signal Strength Increase.
3. SSH. This is the level that produces a request for a handoff (Signal Strength Handoff).
4. SSB. In certain conditions signals may travel further than desired and cause interference in other clusters. If a voice channel is not being used but its received level exceeds the SSB (Signal Strength Blocking) it can be blocked and taken out of service to avoid the interference.

Power Levels

The power levels of a class 1 and 2 mobile phone on the TACS system can be controlled from the mobile switching centre (MSC) via the base station. This ensures that only the necessary power is radiated in order to maintain a good quality of signal. Excessive radiation of power is liable to cause interference. Class 1 mobiles have eight levels of power available and selection is controlled by a three-digit binary code (VMAC – Voice Mobile Attenuation Code). The code is sent over the forward control channel or the forward voice channel and is adjusted as the mobile moves nearer to or further from the base station. The power of a class 1 mobile can range from nominally 7 W to 4.5 mW in eight stages and each level is identified by a unique code of three digits.

In classes 2, 3 and 4 the maximum power is much lower than in class 1. The nominal power levels are:

Class 2: 2.8 W
Class 3: 1.1 W
Class 4: 450 mW

Selective handoff

The Cellnet network uses intelligent base stations to achieve handoff between the base stations by measuring the signal strengths of the voice channel. As a mobile moves away from a particular base station and the mobile's power is increased, the point is reached where the power can be reduced by a handoff to another base station.

Traffic can also be transferred to base stations with spare capacity from a cell with heavy traffic. This is achieved by setting a higher SSH value in the high traffic cell to that in an adjacent cell. The busy cell, therefore, requires to receive a higher power from the mobile, possibly associated with being nearer the base station, in order to stay with the cell. However, to work with the new cell may mean increasing the mobile's power and the possibility of co-channel interference must be taken into consideration. It is, therefore, more advantageous to keep the mobile operating within the correct cell whenever possible.

Mobile identification number

It is an important part of the system that a mobile phone is recognised as a legitimate user of the system. It is also a requirement of the system that the subscriber can be located as, unlike a fixed phone subscriber, the user can be anywhere on the system and links have to be established for each call.

To provide this information a series of numbers are used, some of which are fixed in a PROM which remains with the mobile unit. The numbers identify the phone with a telephone number, the operating network (Cellnet, Vodafone) and the country code (UK).

The manufacturer also provides an 11-digit identification serial number which represents the manufacturer's type and model number. This is placed within the receiver during manufacture and is transmitted when a call is made. If the number is changed the user will be unable to gain access to the network.

A cellular phone may be either fixed power or, as previously explained, capable of operating with various power outputs controlled by the network. The class of operation (1–4) together with the number of channels it can access are registered by the phone together with the mode of operation.

An important part of the system is the home location record. This is information which is stored on a computer and is accessed whenever the subscriber is called or registers within an area. It identifies the subscriber and indicates the part of the country in which the subscriber is at present or is normally located. Local paging, rather than nation-wide paging, can then be used when a subscriber is called.

Registration of the mobile

In order for the mobile to be located and calls to be routed by the area mobile switching centre (MSC), it is necessary for the mobile to register when it is in an area. The MSC must also become aware of mobiles which are about to enter its area. Registration takes place in the following circumstances:

1. The subscriber wishes to make a call and accesses the system.
2. The MSC commands the mobile to register.
3. When moving across the boundary between the area covered by one MSC and another.
4. At predetermined times. This is automatic and requires no operation by the subscriber.
5. When the mobile has been switched off, it immediately registers when it is switched on.

Measuring channel signal strengths

Before handoff of a moving mobile from one cell to an adjacent cell can be performed, it is necessary for the MSC to be aware of the signal strengths received from the mobile in adjacent cells. This is accomplished by each cell being equipped with a dedicated receiver (signal strength receiver – SSR) and control unit whose function is to measure the signal strengths of all the mobile frequencies of adjacent cells on a cyclic sampling basis. The results are stored in the control unit and the mean value of each sampling in all the cells is compiled so that each cell is aware of the transmission quality available in adjacent cells with every mobile. The MSC is, therefore, aware which cell has the best transmission quality when the handoff has to take place. A new voice channel is selected and prepared in the new cell. If all the channels are in use in the preferred cell then a second choice cell is selected. Change-over between the cells requires the mobile to be informed of the new channel frequency and the new SAT (Supervisory Audio Tone) frequency associated with the new cell. A signalling tone (ST) is used to synchronise the change-over both in the MSC, to switch the routing, and in the mobile, to change the channel frequency.

The change-over procedure can be summarised as follows:

1. The signal strength from the mobile reduces as the mobile moves away from the base station. This is detected by the MSC and initially the mobile is instructed to increase power.

2. The signal strengths received in adjacent cells from the mobile are sent to the MSC and the best cell is selected.

3. The MSC selects a new cell, voice channel and SAT and the mobile is informed.

4. The MSC sends a command for the mobile to handoff and a synchronisation message is returned to the MSC.

5. The mobile retunes to the new cell and channel and loops the SAT back to the MSC.

6. The MSC detects the new SAT and that the change-over has been successful. This allows the call to proceed with the minimum disturbance.

When a mobile is paged the information is passed to all the cells in the particular area. As the number of subscribers on one MSC increases and reaches approximately 50 000 there may be difficulties in paging and it becomes advantageous to establish location areas. When this system is in operation and the mobile moves from one cell to another it immediately re-registers its new position to assist future paging. This is achieved automatically by a RAM (Random Access Memory), fitted within the mobile, keeping the last area identification code (AID) within its memory. When the mobile crosses the boundary it receives a new data signal containing the new AID within an overhead message (see later in the chapter). This is compared with the AID in memory and, if different, a new registration is made.

The cellular mobile network

Since the inception of cellular radio in the UK the networks have rapidly increased physically and in complexity. As the number of subscribers has risen, the cell size has been reduced, allowing greater frequency reuse, and the number of channels has increased with ETACS. The network to the base station is fully digital and analogue signals are only used between the base station and the mobile. As in all telecommunication networks, the majority of the system is involved in routing, signalling, charging and supervising the call. The actual transmission of speech is only a small part of the system. The cellular network, however, is made more complex than a normal PSTN due to the necessity to provide for the mobility of the subscriber. Most calls to or from a mobile phone either originate or terminate on an office or home phone and the cellular network must, therefore, be fully integrated into both the BT and Mercury PSTN networks.

Although the basic principles and features of the Cellnet and Vodafone networks are similar, different manufacturers' equipment has been used to construct the networks and, therefore, there are differences in the engineering methods used to provide the service. Both systems, however, have the same problems to solve and similar solutions have been adopted.

Databases must be provided at the mobile switching centres to store all the information about the subscribers. The mobile's telephone number contains digits which indicate the location of the subscriber's home register. This can be the local MSC or a more remote MSC to which the subscriber was allocated when the phone was

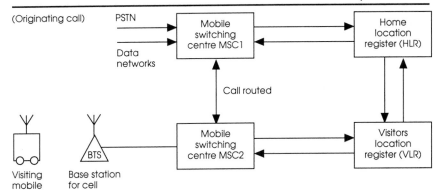

Figure 10.10 *Home and visitors location registers (Vodafone). When call is received from PSTN MSC1 interrogates HLR to verify subscriber and locate mobile for paging. Master information files on all mobiles are registered at MSC1. Mobile's information is transferred and stored in VLR on the switch serving the cells that mobile is visiting. The mobile periodically registers its location and that it is active.*

registered with the network. This is accessed each time a phone is switched on and the information is transferred to the local area in which the mobile is presently situated. The database in which the information is stored is known as the home location register (HLR) and, when accessed, the information about the individual subscriber is transferred to the MSC serving the cells where the subscriber is located and, on the Vodafone network, is stored in a visitors location register (VLR). The current VLR for any subscriber is recorded in the HLR. Cellnet does not use a VLR and only interrogates the HLR.

When a subscriber receives a call, the telephone number allows the HLR to be interrogated and, if the phone is active, the present VLR is known and the subscriber can be locally paged. All the cells in the area where the mobile is currently located are paged on the control channel.

Mobiles are programmed to re-register approximately every 15 minutes to establish that they are still active. If this does not occur, due to the phone being switched off or the mobile losing the signal, the phone is classified as inactive and is not paged should there be an incoming call.

Should the mobile phone be inactive for any reason, the system is capable of rerouting the call to either an office or a home number, or to a network voice messaging facility. This facility can call the subscriber on a paging system or the message can be automatically replayed when the mobile phone again becomes active.

Field Strength Testing

As the networks expand and cells become smaller, the economics of only site testing to establish the field strengths necessary to ensure the correct coverage and propagation characteristics for a high quality system to the required 90% of the UK population becomes prohibitive. Propagation models have, therefore, been constructed and it has been found that the predicted coverage from a particular site has been sufficiently accurate to allow the use of these programs in the construction of the TACS network.

Exchanges

The network was initially constructed with digital switching centres (MSCs) spread throughout the country and linked by BT or Mercury lines. As the number of mobile switching centres increased, an overlay system was added by the installation of a number of transit switching centres (TSCs). Each MSC is now connected to at least two TSCs to provide security for the national network. This has reduced link costs and made the system more manageable. The Vodafone system at the time of writing consists of 20 MSCs, 10 TSCs and 700 base stations spread throughout the UK.

Subscribers are normally connected to offices and homes by the PSTN but where an organisation has a large number of mobile phones in operation it can be more economic to provide a private line between the MSC and the company's offices.

As previously explained, digital exchanges consist of a number of sub-systems each of which performs a specific function in the operation of the digital telephone exchange. This system allows a variety of exchanges to be built by the interconnection of the sub-systems. The digital systems operate under the control of software and it is the programs that create the versatility of the systems and the method of operation. In the Vodafone network Ericsson AXE 10 exchanges are used while Cellnet uses exchanges manufactured by Motorola.

Call procedure on the cellular network

Outgoing call to the mobile

1. A call to a mobile is routed to the mobile switching centre (MSC) from the normal telephone network (PSTN).
2. The home location register is interrogated using the telephone number of the subscriber. It is confirmed that it is a valid subscriber, the phone is active and the present location is obtained.
3. The mobile is paged on the control channels to the cells in the area in which it is last registered. After being paged the mobile rescans all the control channels to find the strongest control channel signal.
4. The mobile responds on the control channel to the MSC.
5. The MSC selects a free voice channel, switches on the voice channel transmitter in the base station and gives the mobile, on the control channel, both the channel frequency and the frequency of the supervisory audio tone (SAT).
6. The mobile tunes into the voice channel frequency and loops the SAT back to the MSC.
7. The MSC instructs the mobile to turn on the signalling tone (ST) and ring the subscriber.
8. The ST is also returned to the MSC.
9. The subscriber lifts the handset and the ST is interrupted at the MSC.
10. The call is connected by the MSC to the mobile via the base station and the conversation can commence.

11.When a channel is not in use the channel transmitter is switched off. This reduces unnecessary interference. A free channel is detected when no carrier is received.

Mobile originated call

The mobile is located in a particular cell served by a base station. The base station is connected back to the MSC by a permanent link. A number of voice channels and a control channel are allocated to the cell. The mobile is on standby on the control channel. The sequence of operation is as follows:

1. The mobile subscriber dials the required telephone number and pushes the SEND button.
2. The mobile unit transmits both the identification number and the called number on the control channel. The MSC uses the identification number to check the customer information in the visitors location register. If the mobile is registered, the number the subscriber requires is called.
3. The MSC selects a voice channel frequency, informs the mobile of the frequency and SAT and starts the transmitter in the base station.
4. The mobile tunes into the channel and receives the SAT.
5. The mobile checks it is the correct SAT and loops it back to the MSC on the voice channel.
6. The MSC receives the returned SAT and calls the required number.
7. The mobile can hear the ringing tone.
8. When the called number answers the conversation can commence.
9. The SAT frequency is continuously looped back to the MSC for the duration of the call.
10. When the call is completed the subscriber hangs up and a signalling tone is sent from the mobile to the MSC for 1.8 seconds.
11. The signalling tone is detected by the MSC and the call is released. The voice transmitter at the base station is switched off.
12. The mobile switches off its own transmitter and remains tuned to the control channel.

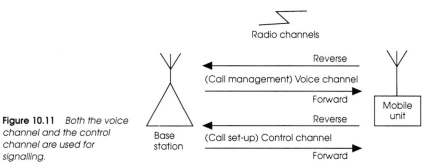

Figure 10.11 *Both the voice channel and the control channel are used for signalling.*

Signalling

The network relies on a complex signalling system in order to establish the links, validate the subscriber, switch, monitor and register the mobiles. In order to achieve these there is a control channel to each cell through which all active mobiles are in regular communication with the mobile switching centre (MSC). The control channel consists of a forward control channel (FCC) from the base station to the mobile and a reverse control channel (RCC) from the mobile to the base station. Data is transmitted at an 8 kbit/s rate using binary code (1, 0) and frequency shift keying (FSK) as the modulation format (see Chapter 12).

Frequency modulation is used on the voice channels and the speech has a deviation of 5.7 kHz with a maximum deviation of 9.5 kHz.

Two audio tones are transmitted in the forward and reverse directions on the voice channels. One is the SAT frequency which is sent from the MSC and looped back by the mobile for the period it is operating on a channel in a particular cell. When the mobile changes frequency and cell a new SAT is received and looped back as previously explained. The SAT is one of three frequencies and is transmitted as an FM signal with a deviation of approximately 1.7 kHz. Although the SAT is sent continuously the speech is not affected as the SAT is a much higher frequency than the band of speech frequencies.

The second tone is a signalling tone and is at 8 kHz ±1 kHz. This is sent as an FM signal with an approximate deviation of 6.4 kHz. It is only sent in the reverse direction from the mobile. This is used for four purposes and the differences for which it is used are determined by the duration of the signal.

1. When the mobile receives a command to hand off to a different channel and cell, the information relating to the new channel frequency, SAT and power level is stored by the mobile and a confirmation ST is sent for 50 ms on the reverse voice channel (RVC) before handing off.
2. During a conversation the subscriber may wish to request other services and this is indicated by a Hookflash. In practice this is a burst of ST over the RVC for 400 ms.
3. When the call is terminated, this is indicated by an ST on the RVC for 1.8 s.
4. Once a mobile has been paged and alerted, the ST is returned on the RVC for a maximum period of 65 s or until the call is previously answered by the subscriber.

Data signalling

Data signals are sent on the forward control channel from the MSC and the reverse control channel from the mobile. The messages are sent as words and each word consists of 28 bits plus 12 error correcting bits.

Messages begin with synchronisation bits followed by two words. Each word is repeated five times to ensure the message is secure. As there is usually only one control circuit for each cell it is essential that messages from the mobiles do not collide when they are entered, and to avoid this a Busy/Idle bit is inserted after bit and word synchronisation and every tenth message bit. Mobiles monitoring the bit change are able to send

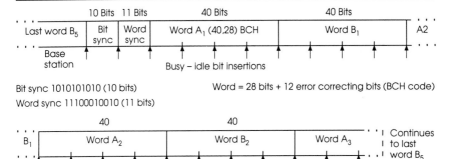

Figure 10.12 *Forward control channel (FCC). Each word is repeated five times ($A_1 - A_5$) ($B_1 - B_5$) after bit and word sync. After every 10 message bits the Busy–Idle bit is transmitted to indicate whether reverse control channel is free. Control channel used by all mobiles and messages can only be sent when the channel is idle.*

their messages when the RCC is clear. As soon as a mobile initiates a call the bit is changed to Busy and other mobiles cannot send until the bit changes to Idle mode.

Messages to Mobiles

All messages are addressed to an individual mobile and contain the mobile station identity. The important messages are:

1. Page. The mobile identification is given by 10 digits.

2. Voice channel designation. This is an order for the mobile to switch over to a particular voice channel selected by the MSC for the conversation. A two-digit code (SCC – Supervisory Audio Tone Colour Code) indicates which one of the three SAT frequencies is to be transmitted on the channel. A further code indicates the output power which is to be used on the channel.

3. Directed retry. If during the call set-up procedure the MSC is unable to select a voice channel in the cell because all channels are busy then the control channels in six adjacent cells are indicated. The mobile will select the best from amongst the six and contact the MSC. The MSC then continues the set-up using the new cell and finds an available speech channel.

Overhead messages

The mobile station equipment has been designed to function with different systems and different operation procedures. The system overhead message is, therefore, transmitted approximately every second to inform the mobile about the operating parameters of the system. The message is received by all mobiles. The messages in digital form carry information relating to the:

Area identification
Type of system
Identification of the group of cells
Number of paging channels in operation

Number of access channels in operation

Number of digits to be sent for mobile station number

Indication if periodic registration required

Provision of priority for subscribers

Prevention of overload on reverse control channel (RCC) (access prevented to RCC for short periods)

Control filler message when no other message being sent on FCC

When messages need to be sent they are always transmitted in a specific order and any spare time is filled with the filler message.

Many acronyms are used for the various messages and all the above messages are included in four groups of overhead messages:

SPOM – System Parameter Overhead Message

GAOM – Global Action Overhead Message

REGID – Registration Identification Overhead Message

CF – Control Filler Overhead Message

Reverse control channel

The mobile communicates with the base station and MSC on the reverse control channel. Synchronisation bits and seven digits (DCC – Digital Colour Code) indicating the group of cells transmitting precede each frame. Each word contains 36 bits and 12 parity bits and is repeated five times for security of message.

The messages sent can be in response to paging calls or dialled calls can be made to the MSC. The mobile provides its identification and serial number, its type of equipment and the maximum output power.

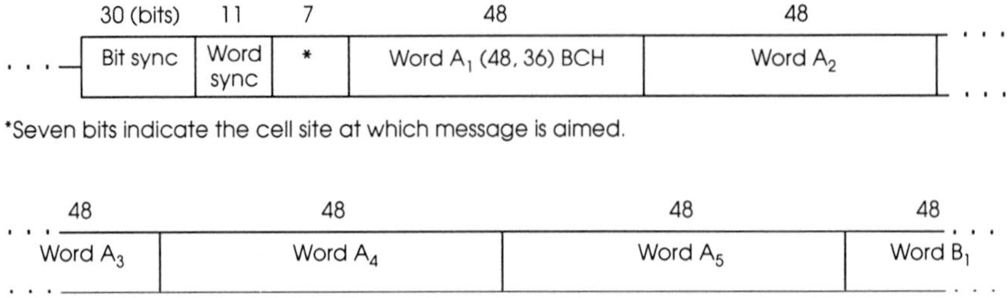

30 (bits)	11	7	48	48	
Bit sync	Word sync	*	Word A₁ (48, 36) BCH	Word A₂	

*Seven bits indicate the cell site at which message is aimed.

48	48	48	48	
Word A₃	Word A₄	Word A₅	Word B₁	

Figure 10.13 *Reverse control channel (RCC). Each word contains 36 message bits and 12 parity bits for error correction. Each word is repeated five times. Reverse control channel is shared by mobiles and can only be used when idle.*

Data on the voice channels

During a handoff digital data is transmitted on the forward voice channel (FVC) to inform the mobile of the new voice channel and new SAT associated with the new cell. The mobile will also be told the output power required for the new cell.

Should the output power of the mobile need to be changed during communication the order is passed from the MSC over the FVC as also is the information to release a call.

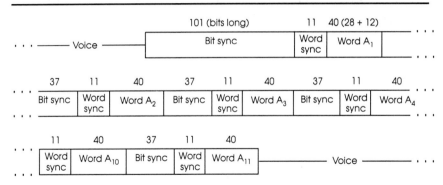

Figure 10.14 *Forward voice channel (FVC). Each word consists of 28 message bits and 12 parity bits. The 28 message bits are arranged into groups to represent different information.*

There are 101 synchronisation bits preceding each word sync and each word is repeated 11 times for security of message.

Reverse voice channel signalling

The format is similar to that used on the FVC but words are only repeated five times. It is used to confirm an order or to call an address for a conference call.

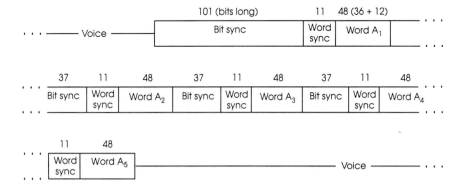

Figure 10.15 *Reverse voice channel (RVC). A word consists of 36 bits of message and 12 parity bits. The 36 bits are grouped to provide individual pieces of information, e.g. 4 bits represent one digit when calling an address. Each word is repeated five times.*

GSM (Groupe Spécial Mobile) – European digital cellular radio (also known as Global System for Mobile Communication)

The second generation of cellular radio has been designed to provide a common standard throughout Europe, with the ultimate advantage to the subscriber that roaming throughout the countries using the system will be possible, and the quality of speech and services that are possible when using a digital system are provided. Digital equipment, because it can be built as common sub-systems and can be arranged in various permutations to build a variety of systems, is eventually cheaper than analogue circuitry. It is also easier to maintain and once the system is working requires very little routine maintenance because it only transmits digital pulses of 1 or 0 rather than the complex waveforms of analogue signals. Complex coding and error correction techniques are used to improve the reliability and bit reduction techniques are used to produce codes capable of providing acceptable speech with 13 kbit/s sampling rather than the 64 kbit/s used by BT in the PSTN (Chapter 7). The many advantages of digital systems are gained at the expense of a more complicated system.

The concept of GSM began in 1982 with the formation of the GSM committee by the Conference of European Posts and Telecommunications (CEPT) and the basic agreement of a system was made in 1990. It was hoped that the specification could be adopted as a world digital standard for mobile communication although international politics and commercial interests may prevent this happening.

GSM radio system architecture

The system is based on a series of contiguous radio cells which provide complete cover of the desired area. Each cell is allocated frequencies but the modulation in GSM is digital and is based on time division multiple access (TDMA) whereby a

Figure 10.16   *GSM phone (Courtesy Cellnet).*

Figure 10.17 *GSM mobile phones (Courtesy Cellnet.)*

Figure 10.18 *GSM phones (Courtesy Cellnet.)*

single carrier frequency can carry several channels by allocating time slots to each channel (Chapter 7). (The UK TACS analogue system uses frequency division multiplexing access (FDMA) whereby each channel modulates and occupies one channel frequency.)

Each cell has a base transceiver station (BTS) which transmits and receives on the cells' allocated frequencies. A base station controller (BSC) operates with a group of BTSs and is responsible for handovers between the BTSs and the transmission power levels. The main control is the mobile switching centre (MSC) and this is responsible for the management of the telephone traffic. This includes:

Call set–up
Routing
Termination
Charging and accounting information

It is the interface between the GSM and the PSTN networks for both telephony and data.

Roaming

The GSM mobile receiver/transmitter or station (MS) ranges between five different classes. These include a class 1 vehicle or portable mobile radiating 20 W to a hand-held portable unit radiating 0.8 W. These units can be fitted with a subscriber

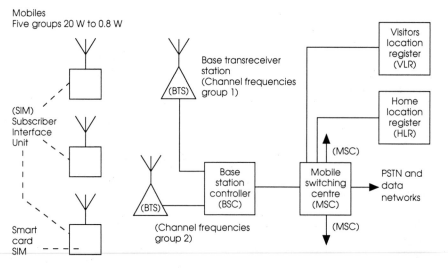

Figure 10.19 *GSM network. The BSC is responsible for handovers between BTSs and transmission power levels. The MSC is responsible for call management, call set-up, routing and termination, charging and accounting information, interface with PSTN and data networks.*

identification interface module (SIM) which can be internal or a slide-in smart card which holds all the relevant details about the customer and the services to which the customer subscribes.

The information is kept on a home location register (HLR) and at least one of these is part of every GSM network. When the mobile is switched on an interrogation takes place between the SIM and the MSC to prove its validity for use on the system and find the location of the mobile. The subscriber's international mobile subscriber identity (IMSI) is used by the local MSC to interrogate the subscriber's home location register (HLR), irrespective of where it is, and a challenge and response confirm the SIM. The relevant information is transferred to the visitors location register (VLR) and a further challenge is made to the mobile SIM. If the response is correct, the mobile is accepted by the network.

Advantages of TDMA for GSM

The spectral efficiency of a digital system is greater than analogue systems. This means that more users can be accommodated within a given area for a given bandwidth. This is achieved in two ways:

1. A low bit rate is used for encoding the audio signals on digital mobile radio and this reduces the bandwidth required by each user compared to digitising speech for the PSTN (Chapter 7).
2. Digital processing can compensate for interference and signal dispersion. This leads to a reduction in the distance before frequencies can be reused and smaller cells are possible. This provides increased capacity.

The carrier bandwidth is 200 kHz and, by using TDMA, eight channels are multiplexed on each carrier. TDMA, however, needs precise synchronisation in order for the signals to be meaningful and this adds to the complexity of GSM. It becomes even a greater problem when the mobile is moving. This causes the time for the radio signal to pass between the mobile and the BTS to be constantly changing. However, with correct synchronisation, there are benefits to TDMA coding:

1. Unused transmission and reception time slots can be used as control channels and these provide additional control facilities which improve the system. Each mobile monitors up to16 surrounding BTSs on a dedicated broadcast control channel and places in memory the six best receptions. This list is used for fast handovers as the user moves between BTSs.
2. Because only one-eighth of the transmission period is occupied by one mobile, there can be a saving in the battery power used by the mobile.
3. As there are eight channels modulating each carrier there is a saving in transceiver equipment compared to the analogue systems where one carrier is modulated by one channel.

The number of channels transmitted on each carrier is limited by the bit rate. At the frequencies used, reflections cause multipath transmissions. The combined signals interfere with the pulses and make their identification at the receiver difficult to decode. The higher bit rates produce shorter pulses and interference caused by different paths becomes more apparent. The specification requires the receiver to equalise the distortion caused by the multipath reception for a time difference equal to 4-bit periods. Eight channels are, therefore, a compromise and require a bit rate of 270.833 kbits/s (Figure 2.3, page 19).

Multipath transmissions also cause fading where the direct and indirect signals are of opposite phase due to the difference in distances travelled. Half a wavelength at 900 MHz, which causes opposite phased signals, is only 16.6 cm. The bandwidth, however, of a TDMA signal is wide and only part of the signal is affected at any time. Fading is, therefore, less of a problem with TDMA and the system can function with greater interference providing the receiver can equalise to compensate for the distortion.

Digital speech and channel coding

Once a signal has been encoded in a digital format, the bits do not have to be transmitted in a sequential manner. There are advantages to redistributing the pulses so that impulsive interference does not affect a complete block of sequential information. Only a very small portion of the original signal of any particular channel is affected due to the redistribution. These techniques have evolved to take advantage of the fact that the signal comprises pulses representing only a minute part of the signal and are time related. Providing the digital signal can be restored to the correct order before demodulation then the techniques help to provide a reliable and an accurate reproduction of the original signal. GSM takes advantage of these coding methods.

The method used by GSM is as follows:

1. Quantisation of the speech (Chapter 7) produces, in a digital format, encoded speech of 13 kbits/s. This is arranged in 20 ms blocks of 260 bits (13 000 × 20/1000 = 260).
2. The 260 bits are split into those representing the 182 most significant bits (MSBs) and the 78 least significant bits (LSBs). Additional bits are added to the MSBs to provide error detection and forward error correction (FEC); 196 bits are added to the MSB to produce 378 plus the 78 original LSBs to make 456 bits for each block. Digital systems allow bits to be added depending upon the transmitted data and comparisons at the receiver allow corrections and errors to be detected by complex coding systems.
3. The 456 bits are split into eight sub-blocks of 57 bits each. Before transmission these are interleaved and redistributed within the time slots (item 5).
4. The GSM waveform consists of a succession of multiframes each of which contains 26 frames. Frames 13 and 26 are used for control and signalling and the remaining 24 frames are used for the traffic.
5. Each frame is divided into eight time slots. The 456 bits in a block (item 2) require the time of four time slots but with interleaving and redistribution they are spread over eight time slots and eight frames. Each time slot accommodates 57 bits of the block (456/8 = 57). The spreading of the bits throughout the frames reduces the effects of impulsive noise.

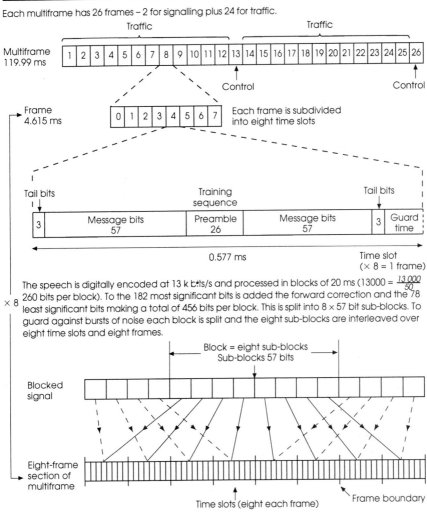

Each multiframe has 26 frames – 2 for signalling plus 24 for traffic.

The speech is digitally encoded at 13 k bits/s and processed in blocks of 20 ms ($13000 = \frac{13\,000}{50}$ 260 bits per block). To the 182 most significant bits is added the forward correction and the 78 least significant bits making a total of 456 bits per block. This is split into 8 × 57 bit sub-blocks. To guard against bursts of noise each block is split and the eight sub-blocks are interleaved over eight time slots and eight frames.

In the hierarchical frame structure of GSM, a multiframe is part of a superframe, which is part of a hyperframe
   1 hyperframe = 2048 superframes. Time 3h 28min 53s 760ms (see page 196).
   1 superframe = 51 multiframes (each has 26 frames). Time 6.12s (shown above).
      or = 26 multiframes (each has 51 frames). This carries different signals.

Figure 10.20   *GSM multiframe and interleaving of data.*

6. Additional protection of the data is supplied by changing the transmission frequency of the channel if the quality is reduced on the original frequency. Frequency hopping can take place once per frame ($1000/4.615 = 217$ times/s).

Method of equalisation

As previously explained a transmission signal to and from a moving mobile is subjected to a changing pattern of multipath transmission. The receivers must be continuously updated with the information for equalisation and this is performed every time slot. A 26-bit pattern (training sequence) is transmitted every time slot and at the receiver it is compared with a standard pattern. The difference is used to modify the digital equalisers to correct the signal parameters.

Advantages of GSM

In recent times there has been concern that the analogue TACS system is susceptible to eavesdropping with the modern scanners available in the shops today. Such listening systems require both transmission and reception frequencies to be tuned into in order to hear both halves of the conversation. The GSM digital system, however, provides a complex encryption to the data bits which requires an algorithm to decode the data. Only the participants are, therefore, able to receive the decoded data.

The algorithm that calculates the ciphering key uses the TDMA frame number as one of the input parameters. For this purpose a cyclic period of approximately 3.5 hours (2 715 648 TDMA frames) is used in which all the frames are sequentially numbered before being repeated as the frame numbering cannot continue to infinity.

At the time the subscriber joined the network a subscriber authentication key (K_i) was issued together with an international mobile subscriber identity (IMSI). The key is stored at the authentication centre on a database and in the SIM of the mobile. At the Authentication Centre it is used, together with a random number and two algorithms, to produce three pieces of data for the system. These are known as triplets and are (1) ciphering key K_c, (2) signed response for the system, (3) random number. These are stored in the home register.

The signed response is used when authentication is required and this takes place at each registration, each call set-up, location updating and before supplementary

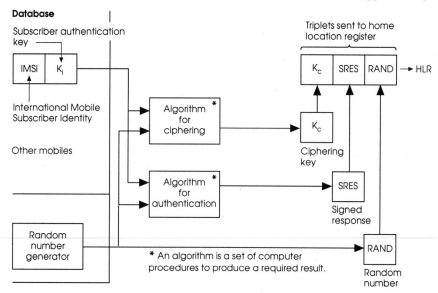

Figure 10.21   *Authentication centre.*

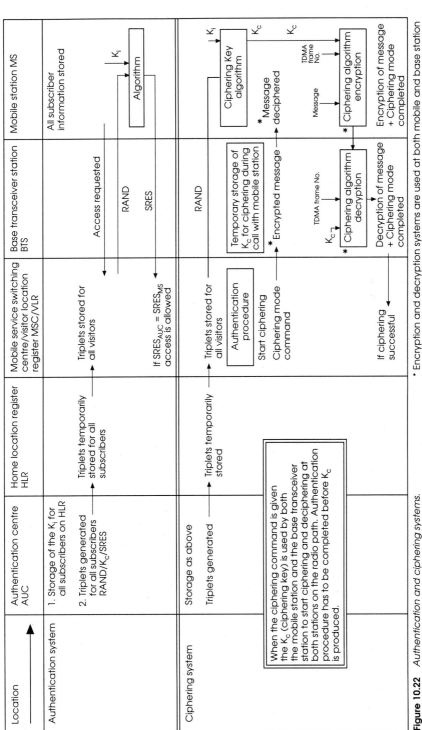

Figure 10.22 *Authentication and ciphering systems.*

* Encryption and decryption systems are used at both mobile and base station

services are provided. Each time the response from the mobile has to be correct otherwise the mobile is not accepted.

The simplified figures show the signals required to provide the encryption and decryption for the system.

The main advantage to the system designer is that the digital system provides increased spectral efficiency. This is achieved by the fact that GSM is able to operate with carrier/interference ratios of 10–12 dB compared with 17–18 dB necessary for the TACS analogue system. The additional interference that can be tolerated, due to the signal processing, allows the available frequencies to be reused at much closer distances. This provides efficiencies which are up to three times greater and this may further improve with bit rate reduction systems in the future. In both GSM and TACS a 25 kHz bandwidth is requred to transmit one channel.

Processing power requirements

The GSM system is far more complex than the analogue TACS systems and a considerable amount of the data (approximately 62%) is used for signalling, control and error correction. The remaining 38% of the data is used for the basic communication. The speech data on each carrier is 104 kbits (8 channels × 13 kbits/s). The total data rate with the additional bits added is 270.833 kbits/s. This requires considerable processing power at both the base station and mobile. Dedicated microprocessors, however, are allowing equipment to be the same size and power consumption as present day TACS equipment and with the possibility of considerable improvements in later generations of equipment.

Vodafone and Cellnet will provide a national coverage for the GSM system in addition to the TACS networks. The system has been approved not only in Europe but by countries in Africa, the Gulf states, Australia, Singapore, Malaysia, Thailand, Taiwan, New Zealand, Hong Kong, Sri Lanka and Macau. Japan and America are producing other digital cellular systems and will provide competition to GSM especially in the Pacific and Asia.

Personal Communication Network – PCN

The small pocket telephone capable of being used anywhere and having the capability of communicating both nationally and internationally is the ultimate system for telephony. Such systems will operate with cellular networks but the desired size affects the possible power of the equipment and the potential number of users makes channel planning and spectrum efficiency major problems. Such systems are, however, planned to become national networks and operational systems commenced in 1993 in the London area, and are being expanded.

The system adopted is based on the ETSI Standard and it is intended that the networks throughout Europe will be harmonised both for mobility of use by subscribers and scale of production for manufacturers. This allows economic systems and equipment to be produced.

The system adopted by the PCN licence applicants is the GSM Pan–European Mobile System but adapted for use in the allocated frequency band of 1710 MHz to 1880 MHz. GSM already defines the radio air interface and the network. This includes specifications for the radio access, mobility, switching and transmission. The new standard for PCN based on GSM is known as DCS 1800 (Digital Communication System 1800 MHz).

To date the TACS cellular networks have concentrated on the business user for their customers but PCN is intended for a mass market for both speech and data and will eventually compete with the conventional PSTN both at home and in business.

In order to provide the service required the PCN system must be capable of:

1. A wide area coverage which is continuous and reliable.
2. A quality of speech and operational reliability.
3. Working with a small lightweight portable phone.
4. Using the channels efficiently and providing a good grade of service.
5. Communicating at speed within an area covered by the network.

The frequencies allocated to the PCN service produce their own problems. The high frequencies mean that the cell size is reduced in order to provide reliable service. This can be an advantage as it increases frequency reuse and makes available more channels. However, it requires a greater number of sites, in order to provide base stations, and these are difficult to find. The high frequencies of 1800 MHz may also provide propagation difficulties due the the size of a wavelength ($3 \times 10^8/1.8 \times 10^9 = 0.166$ m). High attenuation can be associated with such small wavelengths by obstacles and even weather although penetration through windows and into buildings is good.

Contiguous coverage depends upon the height of the base station antenna. If it is situated below roof-top height the signal is confined by the buildings to down the length of streets. Although this would provide microcells of high capacity for

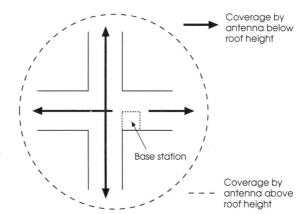

Figure 10.23 *If base station antenna is below the roof height the radiation is restricted to the roads by the buildings. If the antenna is above roof height an omni-directional coverage is achieved.*

particular areas, it is not economic and high antennae are provided which are situated above the roof tops.

Cell size

The cost of the network depends on the number of base stations that have to be installed including the link by cable or radio of the base station to the mobile switching centre. The smaller the cells, therefore, the greater the cost. The cell size is affected by:

1. The mobile and base station transmitter output power and radiation pattern.
2. Multipath propagation. This can cause fast fading which, unless complex engineering solutions are provided, reduces the range.
3. Multipath signals arriving at the receiver. These can cause delay spreads which become significant when the bit energy is spread over a period comparable with the bit period. If the antenna is situated above roof height this causes more delay than a lower antenna. To compensate for the delays a channel equaliser is provided in the receiver.

The channel equaliser introduces complexity in both the base station and the receiver. It uses the training data in the GSM time slot and compares the received signal with the expected signal and calculates the digital filter parameters necessary to restore the signal (see Figure 10.20).

When the mobile is in motion and it is receiving direct line-of-sight signals it is still subject to a frequency change due to the Doppler effect. (Frequency increases if the mobile moves towards the base station or decreases if it moves away. Frequency shift depends upon the speed of the mobile – principle used in police radar speed guns and burglar alarms.) In a situation where multiple path reception is experienced, the Doppler effect causes a variable spread of frequency and when this is comparable to the period of a bit then error rates increase. The equaliser has to compensate for these errors.

Handover

Handover has to take place when the mobile moves between cells. In addition the handover technique can be used to provide coverage where the radiation propagation is irregular and the mobile is able to receive signals from two transmitters. It may even occur that a moving object such as a lorry or bus temporarily obscures a mobile from a transmitter with which it is communicating. In these circumstances it may be possible to hand over to another transmitter with a direct link.

Efficient use of the radio spectrum

If PCN is to be successful the system must be capable of providing a high traffic capacity although, as in all radio systems, the capacity is ultimately limited by the finite spectrum available. The efficiency with which the bandwidth is used to provide the channels is determined by the method of modulation, the speech coding rate and the carrier spacing. In addition the distance between the frequencies being reused determines the number of channels that can operate in an area. The ability to reuse frequencies depends upon the tolerance of the signal to interference. It must, therefore, have good immunity to co-channel and adjacent channel interference. As PCN is transmitted in digital form, as in GSM, it has been previously shown that much higher levels of interference can be tolerated when compared to analogue systems.

When coding speech into data, the codecs operating with high data rates can provide good quality speech and low delay. However, for a given bandwidth, codecs operating with low data rates provide greater capacity but produce greater delay. Delay, if excessive, is important in such systems as it can be disturbing to the subscriber and requires the echo control to compensate on the network. The speech coding rate must, therefore, be a compromise between the capacity of the system and the quality of the speech. In the same way as GSM, error protection is provided by additional bits for forward error correction checking and by interleaving the data so that impulsive interference does not affect a large area of adjacent signal.

Handset

The necessary compactness and light weight places considerable design restrictions on the handset. In addition, as the telephone is operating in a cellular system the transmitter power determines the operating range of the equipment. The battery is the most important item for determining these characteristics as it is both the largest and heaviest component and its capacity determines the available power. The unit would not be possible without the electronics being in integrated form and it owes its existence to advances in VLSI technology. Included in the handset must be electronics for speech and channel coders, channel equalisers and all the supervisory and control systems for paging, registering, calling, dialling, channel changing, supervision and in addition a receiver and transmitter. Although power is desirable for longer range, it can cause spurious emissions which can seriously affect the network (see the section on multi-operator working). An RF filter can eliminate some of the transmissions but it must also be capable of handling the power. Two mobile power classes have been specified: 250 mW and 1 W. Cell size can vary from 0.5 km in cities to 8 km in rural areas for the 1 W transmission. (In GSM, where 20 W output is possible, the cell size can be as large as 35 km.)

PCN operation with DCS 1800 Standard

The GSM 900 Standard adopted for the European Digital Cellular Networks provided the majority of the requirements necessary for a PCN system but needed adaptations for:

1. Working at 1.8 GHz.
2. Using low power handsets. (PCN is intended for a majority of handportable

users whereas GSM is intended for the majority of mobiles to be in vehicles.)

3. Operating close to a base station which is not the one to which the subscriber is connected (see the section on multi-operator working for an explanation).

4. Increased available bandwidth (GSM 50 MHz in 900 MHz band, PCN 150 MHz in the 1.8 GHz band).

5. Ability to use other PCN networks.

The DCS 1800 specification provides the additional requirements and modifications to the GSM specification. The PCN network uses a split duplex system in which two separate frequencies are used for transmission to and from the base station and mobile. The frequencies are from two bands which are separated by 20 MHz. Each band has a 75 MHz bandwidth (total 150 MHz) and is the largest bandwidth ever allocated for commercial use.

Multi-operator working

The bandwidth is intended to be used by several licensed operators in order to provide competition. This will result in many base stations operating for the various networks. A problem that can then arise relates to the out-of-band frequencies that can be generated by the handportable. A subscriber can be close to another operator's base station but distant from the base station with which the handset is communicating. It will, therefore, be radiating at maximum power. If it generates emissions which include the near base station's frequencies it is picked up by the base station receiver. The frequency it receives may be one on which it is receiving a distant subscriber on its own network but the interference desensitises the receiver and affects the reception of its own subscriber. Specifications relating to spurious emissions from handportables are, therefore, important to the operation of the networks and the manufacture of handportables.

The various network providers may provide systems in different parts of the country but not possess a national network. It will be desirable, however, for a subscriber to be able to use another network in situations where the subscriber's own network does not exist. Sharing networks obviously has an economic advantage to a network provider who can offer a national system without the cost of a national network. This requires engineering problems to be solved, for the situation can arise that once a subscriber has been accepted by another network they may remain on that network even when their own becomes available. If the subscriber is not on their own network they may affect the charges and facilities available to the user. In the overlap areas between two networks the system must allow the subscriber to return to their own network and this is achieved by the system's tracking and paging the mobiles and providing handovers as soon as possible.

The economics of allowing subscribers to use other networks where an agreement has been made between the operators has been taken a stage further and network systems are being built whereby operators share a common physical network. This is known as 'parallel network architecture' and is made possible by the

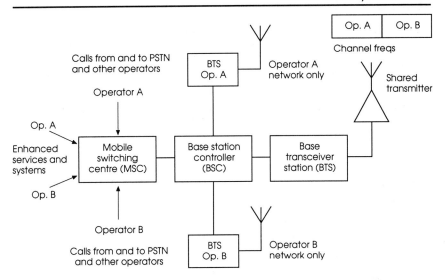

Figure 10.24   *Part of a network can be shared by different operators due to the intelligence designed into the system which allows the individual operator's calls to be separated and transmitted on the operator's allocated channels. These arrangements are known as parallel network architecture.*

intelligence of the system being able to separate and handle channels. In the same way that calls are capable of being routed to different mobiles, so calls can be routed to the relevant network before distribution to the mobiles. Base stations can be added which are exclusive to a particular operator and additional services can be provided independently. The radiated transmissions from a shared base station are split into the different operator's bands. A separate broadcast control channel (BCCH) is provided for each operator's frequency band and this broadcasts the operator's identity. Mobiles are then able to communicate with their own operator and use the correct frequencies.

The block diagram for the GSM network given previously in Figure 10.19 is also appropriate for the PCN network with the use of mobile switching centres (MSCs), base station controllers (BSCs) and base transceiver stations (BTSs). A home location register and a visitors location register are used to identify and locate subscribers.

Cells and frequency reuse

In order to provide the channels for the amount of traffic necessary to justify the PCN systems and also maintain an acceptable grade of service (GOS), the cells are small and the reuse of channel frequencies is high. If channels in a particular area are not available the system becomes blocked and new calls can neither be received nor be originated in the area.

To improve channel reuse two types of cell are used in a layered arrangement. The layer of macrocells is overlaid by microcells in areas of high traffic demand and

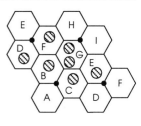

Figure 10.25 *Twelve cells with a 3 x 3 frequency reuse pattern: ABC, DEF, GHI – DEF is shown reused. The macrocells each have 3 x 120 degree sectors and microcells are used in some cells to increase the number of channels. The same group of frequencies is used in each of the microcells.*

● Base stations for macrocells

⊗ Microcell

frequencies are reused in the microcells. The microcells have a very small range and this is achieved by the base station antenna being installed below the roof height which restricts the transmission to a very local area. The same group of frequencies can be used in each microcell.

Base stations for the macrocells radiate into three 120 degree sectors with three separate transmitters, each with a different group of channel frequencies. These same frequencies are not used in adjacent cells which would work with different channel frequencies. Channel frequencies are only reused when co-channel interference does not cause a problem.

Experience with microcells and the development of equipment continues and as the new PCN systems are installed and operated the specifications and engineering can be improved and modified if desired.

A problem arises with small microcells when a vehicle is moving quickly through the area served by the microcell. Calls can be lost, especially at sharp corners, and in these circumstances it is better for the vehicle to be communicating with the macrocell. To distinguish between fast and slow moving mobiles timing mechanism can be used in the mobile which has the effect of reducing the area of the microcell when the mobile first communicates. If after the timing period the mobile is still within the microcell area, the serving area is increased and the mobile selects the microcell for communication. A fast moving mobile would have left the area of the microcell by the end of the timing period and, therefore, stays with the macrocell.

The large number of base stations requires a considerable network of links back to the base station controllers which must carry all the traffic. The most economic way is by radio links and 38 GHz and 55 GHz have been set aside for these transmissions.

There is considerable optimism about the future success of PCN and a large investment is being made by several major companies. If it is successful it could affect the market for the existing TACS and the recently introduced GSM. Although GSM has advantages when compared to TACS its customers could be the same people who at present are operating with TACS. It is, therefore, a very delicate marketing exercise and the marketing pressure to introduce new systems is more likely to come from new companies in mobile communications rather than existing system operators. At the same time development of new systems is continuing. The next 10 years should be very interesting and many marketing managers and chairmen of the companies involved will have sleepless nights.

11 Private mobile radio (PMR)

The growth of PMR has been continuous since the 1950s even though in the 1980s cellular systems became available. The limitations on growth appear to be caused only by the lack of available spectrum and this may cause a future crisis both for the manufacturers of equipment and present and potential users.

The channel spacing is 12.5 kHz for PMR and consideration is being given to proposals for 5 kHz channel spacing in order to improve spectral efficiency. However, with the increase of licences in a crowded spectrum there is the possibility of an unacceptable grade of service in areas where PMR is used intensely.

PMR is used for both speech and data communication between closed groups of users. These may be a company or service (police, fire, etc.) and the system is operated for their private needs. Systems can cover a small area or can utilise several base stations. Some of the facilities may be shared with other PMR users to reduce costs.

Chapter 7 described the possible communication methods between two phones and showed that duplex, semi-duplex and simplex were possibilities and each method altered the way the system can be used. PMR tends to operate in either single frequency simplex or as semi-duplex. These are chosen to reduce the number of radio channels that are used for communication. The systems require a push-to-talk mode of working.

In single frequency simplex only one channel frequency is used for both directions and a user cannot transmit while receiving. If more than one mobile is on the system they will hear both parts of someone else's conversation.

In a semi-duplex system two frequencies are used: one for transmit and one for receive. In such systems one frequency is used by a dispatcher (base station operator) to communicate with all mobile receivers and one common frequency is used by all the mobiles to transmit back to the dispatcher. Mobiles in such systems only hear the dispatcher to whom their receivers are tuned and not the speech from other mobiles.

It is rare for PMR systems to use a full duplex system.

In a push-to-talk (PTT) system the receiver is tuned and biased to receive incoming messages and the transmitter is switched off. It is not made active until the PTT switch is operated. In order that noise is not continuously heard while the receiver monitors the radio frequency a squelch circuit is incorporated into the receiver so that only signals over a predetermined level are heard (Chapter 6).

The simplest PMR system requires only transmitters and receivers to operate on the same frequency for a group of mobiles to maintain communication with each other. However, the environment, transmitter power and aerial radiation patterns limit the area that can be reliably covered with such a system.

For reliable reception and transmission to the mobiles a fixed base station must be used and the site selected to provide the coverage of the required area. The base station is usually situated at a high elevation and an antenna system and mast used which provide the required radiation pattern and signal strengths. These systems operate in semi-duplex and because all the mobiles can hear the dispatcher it is unnecessary to use complex signalling systems to call up a particular mobile. Some discipline is required by users of the system as the mobiles must ensure the channel to the dispatcher is not in use before using it. Tones on the down-link to the mobiles can be used on more sophisticated systems to indicate whether the mobile transmission channel is free.

If the base station transmitter and receiver are connected together it is possible for the mobiles to communicate with each other. This is called TALKTHROUGH and requires the base station transmitter to be continuously operated. Normally all transmitters radiate only when a transmission is required in order to reduce interference with other transmissions and radio noise generally. Permission, therefore, has to be obtained for TALKTHROUGH operation.

TALKTHROUGH can be manually arranged or activated by a mobile transmission when the base station is unattended. In these circumstances some form of signalling is used to prevent unauthorised access to the system.

PMR signalling

The efficiency and facilities offered by a PMR system can be increased with the addition of signalling to the system. This obviously adds complexity to the system engineering and some signalling systems need the approval of the Radiocommunications Agency before use. Without a signalling system the operator uses voice calling in order to contact a particular mobile. The possible signalling systems used include:

1. Continuous tone controlled signalling (CTCS)
2. Dual tone multi-frequency signalling (DTMF)
3. Sequential tone signalling (STS)
4. Digital signalling.

The use of signalling allows codes to be used for:

1. Selective calling whereby individuals and groups can be selectively called by individual codes.

2. Identification of user whereby a code can be sent either automatically or manually for identification and status.

3. Dialling codes for calling numbers in other telephone networks.

4. Interrogating unattended base stations and sending pre-coded messages.

5. Control of base stations in multi-station networks.

6. Control of receiver voting and operation (see later details).

Continuous tone controlled signalling system (CTCSS)

The use of this system requires the permission of the Radiocommunications Agency but is common in PMR systems as it assists in reducing co-channel interference.

The system transmits selected frequencies below 300 Hz. These are below the audible frequency response of the receiver and consequently are not heard by the user. The tones are continuous for the duration of the call. The system is used selectively to call on a single channel and enable base stations to be selectively operated in a TALKTHROUGH mode. Calling is achieved by the squelch circuits of the receivers selectively responding to the different tones transmitted.

The standard frequencies used are shown in Table 11.1 and the user's precise frequency is allocated with the licence.

Table 11.1   *Standard frequencies for continuous tone controlled signalling system (CTCSS)*

67·0	110·9★	146·2	192·8
71·9★	114·8	151·4★	203·5★
77·0	118·8★	156·7	210·7
82·5★	123·0	162·2	218·1
88·5	127·3	167·9	225·7
94·8★	131·8	173·8★	233·6
103·5★	136·5★	179·9	241·8
107·2	141·3	186·2	250·3

★ Initially frequencies are allocated from the indicated 10 frequencies

Sequential tone signalling (STS)

This system is used to enable mobiles to be selectively called, provide a status indication or allow a base station to be selectively operated in a TALKTHROUGH mode.

The audio tones used are within the audible range of the receiver and voice transmission cannot take place while they are being transmitted.

The codes used are generally EEA or ZVEI and consist of five of the standard selective signalling frequencies which are combined to form a coded address (Table 11.2).

Table 11.2   *Standard frequencies for sequential tone signalling (STS)*

(a) EEA system (UK manufacturers)

Digit	Frequency (Hz)	Digit	Frequency (Hz)
G	1055	6	1540
1	1124	7	1640
2	1197	8	1747
3	1275	9	1860
4	1358	0	1981
5	1446	R	2110

Duration of each tone
40 ms ±4 ms
Start interval
100 ms minimum

(b) ZVEI modified system (German manufacturers)

Digit	Frequency (Hz)	Digit	Frequency (Hz)
1	970	7	1670
2	1060	8	1830
3	1160	9	2000
4	1270	0	2200
5	1400	R	2400
6	1530		

Duration of each tone
70 ms ±15 ms
Start interval
140 ms ±15 ms

Dual tone multi-frequency signalling (DTMF) (Figure 11.1)

This system can be used for telephone dialling and operates with private networks and the PSTN. Authorisation is necessary to operate with these systems.

The signalling code uses two simultaneous frequencies which are chosen from the eight allocated to DTMF. The two frequencies are in the higher and lower audio band and simultaneous transmission prevents the chance of voice frequencies imitating the combination and triggering the system.

Digital signalling

Digital techniques (Chapter 7) are responsible for most modern advances in electronics and offer considerable advantages for signalling. They offer better noise immunity, faster signalling, greater accuracy and flexible systems offering more facilities. The digital binary signals can modulate a sub-carrier and be transmitted in or out of the audio band.

Because of the very short duration of time required to transmit a digital code the digital binary signals can be transmitted in short bursts within the audio band together with speech without affecting the intelligibility of the signal.

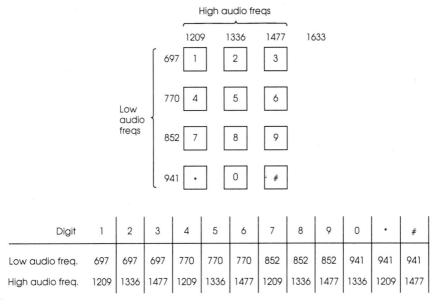

Digit	1	2	3	4	5	6	7	8	9	0	•	#
Low audio freq.	697	697	697	770	770	770	852	852	852	941	941	941
High audio freq.	1209	1336	1477	1209	1336	1477	1209	1336	1477	1336	1209	1477

Each digit is a unique combination of two tones

Figure 11.1 *DTMF – Dual Tone Multiple Frequency signalling. The push button phone sends two tones (one high, one low) to represent each number or symbol. The frequencies shown are common to each row or column. Frequencies chosen are between 700 and 1700 Hz where attenuation and delay distortion are a minimum. Dial tone is also outside this range. The frequencies are specially chosen to ensure there is a minimum risk of digit simulation due to low order harmonics in the set of signal frequencies; 1633 Hz with low frequency combinations are spare.*

The data modulation system uses fast frequency shift keying (see Chapter 12 for explanation).

Digital signalling can be used for:

Selective calling
Status reporting
Pre-coded messages
Vehicle location
Monitoring and supervisory systems
Direct dialling
Trunking (see later)
Mobile terminals including printers and VDUs.

Digital signalling can replace the analogue CTCSS with a low frequency digital signal which can operate the squelch and access repeaters and other facilities with the large number of possible data codes that can be produced. However, at present digital coded squelch signalling is not permitted in the UK and all digital signalling must only be transmitted on exclusive channels.

Control of base stations

The base station may be directly or remotely operated. If it is remotely operated links from the operational point have to be established with it and all normal communication links can be used. These are telephone lines, radio links, fibre optics or direct cable from the operating position when distances are short. In special cases a base station in TALKTHROUGH mode can be operated from a remote mobile.

Control systems for the base station vary considerably depending upon the complexity of the operation. These can range from manual operation to computer and exchange systems. Manual operation is the most widely used system with either voice calling or selective calling. Manual systems can be extended to semi-automatic systems so that while manned the systems work manually and when unattended they work in TALKTHROUGH mode. The operation can be enhanced with the use of signalling codes to set up both group and individual calls and provide the facilities previously listed.

In the larger systems more use is made of automation and several base stations may be in operation and need controlling. There may be many channels operating in different ways with both voice and data being transmitted.

Most PMR systems operate with a single base station and channel. This is only satisfactory when traffic is light. Problems can arise when congestion takes place, especially in the busy hour (Chapter 7), due to an increase in traffic. This problem can only be alleviated with additional channels being provided and this produces the added complexity of channel switching and tuning.

When a single base station is used, coverage is controlled by the antenna height and directional characteristics, transmitter power and the environment. These may lead to unreliable reception in areas that need to be covered. The solution may require additional base stations to serve areas where the signal is unsatisfactory.

The problem is compounded by the difference in the power output of the base stations and small handportables and for this reason a solution may be found by providing additional base stations with only receivers. A system called voting (see later) allows the receiver with the best reception from the mobile to be used while the single base station transmitter communicates with the mobile.

Provision of multi-base stations

Where the reception and transmission from a single base station is inadequate several base stations may be required to cover an area adequately. These systems, when connected to form a PMR network, are known as wide-area systems. The provision of additional base stations and transmitters introduces new problems to the system designer. If two transmitters radiate on the same frequency and both can be received simultaneously, it results in co-channel interference which severely degrades the signal if the signals differ.

Several system design techniques have been adopted to prevent the problem.

Multi-channel systems.

As seen in cellular systems (Chapter 9) co-channel interference is prevented by ensuring that the channel frequency is not reused until the power from the transmitter has been sufficiently attenuated by the environment as to cause no interference in the area in which it is being reused. This requires adjacent transmitters to transmit on different channels. Although this prevents interference it requires a mobile to change channels as it travels through an area. This can be done manually or automatically with modern equipment. The system, however, requires a greater share of the available spectrum which is strictly limited.

Shared channels

There is an obvious advantage for the operation of the mobile if it does not have to switch channels when roaming and receiving its signals from different transmitters. If also the transmitters are on the same frequency there is a saving in the use of the spectrum, which is important for PMR users. However, co-channel interference must be avoided and in a shared system this is achieved by ensuring no two transmitters are operating simultaneously on the same channel. This requires a complex switching arrangement using selective calling and receiver voting. With receiver voting all the outputs received from the mobile on the various receivers are sent back to the control and compared for signal and noise. The receiver with the best quality signal is selected. A method of voting is as follows.

Voting – selecting the working base station

When a mobile calls from within an area served by several base stations, its signal is received by several receivers. Each signal has a different signal-to-noise ratio depending upon the distance of the mobile from the particular receivers. Each base station sends an audio signal to the control position and its frequency is determined by the signal-to-noise value of the mobile to base station link. By this method the control position is able to select the best link to the mobile and reject the other paths. The system is known as the ASSORT system (Automatic Site Selection Of Receiver and Transmitter).

At the receiver of each base station a special circuit, known as the voting circuit, determines the signal-to-noise ratio of the received mobile signal at the output of the discriminator stage. The noise level is then converted to a tone whose frequency is between 2700 and 3000 Hz. The frequency of the tone is proportional to the noise and is generated by a variable frequency oscillator. The speech output at the discriminator is passed through a low pass filter and has a bandwidth of 2.5 kHz. It is then combined with the variable tone and passed to the receiver amplifier stage. In the absence of a received signal the receiver output is reduced by a squelch circuit and noise is not transmitted.

The output of the receiver is transmitted by either land line or radio link to the control position where the ASSORT unit monitors and compares the various noise signals from all the receivers. Signals received are equalised by pre-set attenuators and applied to the input of a logic gate switch. The tones are separated from the signals by filters and are sampled at a 12.5 ms rate by a pulse generated by a clock pulse generator. The sample is passed to a discriminator which produces a voltage proportional to the frequency of the tone. The largest voltage is held in a compare and hold circuit and this relates to the signal with the best signal-to-noise ratio. This signal is accepted by the logic gate. If during the call the situation changes and a different receiver begins to receive the better signal, a change-over can occur which is undetectable to the user.

There is a considerable difference in the power radiated by the mobile and that by the base station. Typically the power can be approximately 30 watts for a base station and 6 watts for a car phone. It can be reasonably assumed that the best path selected from the mobile is also the best return path. The transmitter at the site with the best receiver signal is, therefore, selected to transmit back to the mobile. ASSORT, therefore, automatically chooses the base station transmitter and receiver for the two-way connection.

The signal-to-noise ratio can also be measured at the control position with the added advantage that the noise on the connecting link between the base station and control can be taken into consideration.

A problem can arise with tone voting where under certain propagation conditions continental radio signals can cause sufficient interference that one particular site is always selected by ASSORT. All mobiles are then forced to work to the particular site although in reality it may not be the most suitable.

Manual over-ride is provided in the system and, in circumstances where the associated transmitter with the best selected receiver is not appropriate, an alternative transmitter may be selected. Such situations may arise, on rare occasions, where reflections cause difficulties in the reception of signals at the mobile from the automatically selected transmitter.

Quasi synchronous operation of multiple transmitters

In order to provide satisfactory reception within an area where local topographic conditions may cause reception problems and wide area coverage is not possible, two or more transmitters may have to be used. A system, known as the quasi synchronous system, can be used for transmitter operation with some advantages and disadvantages. The system uses a single frequency for the transmitters instead of different channels to cover the area. The transmitters radiate simultaneously on the same frequency and carry exactly the same modulated information. This form of operation allows a mobile to receive a consistent and reliable signal anywhere within the area.

There are several advantages to the system. All the transmitters can be both modulated and switched from a single point if desired and, when this is combined with the 'voting' control system for receivers previously described, a single channel remote control unit is capable of operating the complex system.

The use of a single channel within the area simplifies the switching arrangement within the mobile unit. The need to change channels as the mobile moves within the area is eliminated and messages avoid being interrupted.

TALKTHROUGH, whereby a mobile can talk to all other mobiles within the area, becomes a simple operation as the receiver needs only to be fed into the transmit line at the control point.

Quasi sync operation provides a considerable saving of channels within an area and, providing individual transmitter coverage is not unusually increased, then channel reuse is not greatly affected.

There is improved reception by the mobile when it is able to receive signals from different transmitters. Signal variations, due to buildings and topography, are reduced due to the increased signal levels and their different directions.

A quasi synchronous system is employed where reception may be difficult but must be reliable and of a consistent quality. The system designer must, therefore, consider many facts. These include the following:

1. Because of the topography and environment there invariably are problems with the variations of reflections and refraction of the signals.
2. Weather conditions can produce variations in the radiated power.
3. The radiation patterns of the transmitters' antennae must be carefully controlled.
4. Consideration of the reception conditions within the area and the type of receivers and antennae being used. Chapter 8 explained the effect of various car mountings and the way the polar diagram and gain is affected.
5. The signals can be subject to long term fading due to the phasing of the various signals.

The system designer must ensure that there is sufficient signal power within the area to overcome these problems. It is also essential that the phasing of the transmitters, the levels and the frequency responses of the signals are correct. In order to achieve accurate phasing of the signals accurate and very stable transmitter oscillators are required and these must be better than those used for a conventional system.

Quasi sync chopping

This is an effect that occurs when two areas overlap and a mobile can receive equal signals from two transmitters. In these circumstances a chopping effect and distortion can occur especially if the mobile is stationary and the signal levels are low. The problem is caused when two transmissions are received which are of equal amplitude and arrive out of phase with each other. If they subtract they can generate bursts of noise. The squelch circuit, in these circumstances, is switched on and off. Multipath reception can reduce the effect. The effect is more apparent at lower frequencies. At UHF, the multipath reflections produce rapid fluctuations with a moving vehicle and the chopping effect is less apparent.

Overlapping areas with multiple transmitters

The number of overlapping areas where signals occur of equal strength depends upon the number of transmitters operating within an area. Figure 11.2 shows the conditions for two and three transmitters. At these points the signals can add, if in

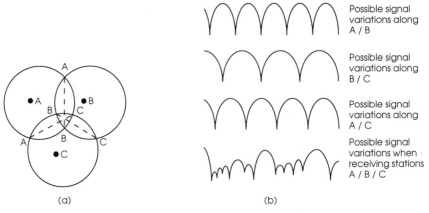

Figure 11.2   *When a mobile is able to receive the same signal from two or more transmitters the signals will interfere with each other. The resulting signal depends upon the instantaneous phases of the different signals at the antenna. (a) Equi-signal points AC AB BC with transmitters A, B, C. (b) Possible variation of signals along equi-signal lines in a three-station system, consisting of transmitters A, B, C. (See also Figure 9.6 paging quasi-sync operation.)*

phase, or subtract, if out of phase, and in the latter circumstances can approach zero signal. Multipath signals and shadows caused by buildings can alter this and equal signal effects along the indicated lines can be random.

The random effect becomes less if the overlap area is increased but this requires either more transmitter sites or higher antennae. In practice more transmitter sites are preferred as more overlapping areas reduce the risk of a complete null of the signal occurring and there is a greater probability of an adequate signal.

Frequency and phasing

The system requires exact frequency and phase synchronisation between the transmitters. If this is not possible there must be controllable differences in the frequencies of the transmitters which are to overlap. In VHF systems the frequency difference is approximately 5-6 Hz. It must be such that beat notes between the transmitter frequencies are not audible. If, however, the frequency difference is too small there are long periods when nulls occur with equal signals which will cause the squelch circuit to chop.

There is less possibility of a complete null with three or more transmitters overlapping. In these circumstances the separation is reduced. Sequential integers such as 2 and 3 or 4 and 3 can be chosen and offsets of 1 Hz and 1.5 Hz (VHF) or 4 Hz and 3 Hz (UHF) can be applied to the transmitters and this reduces the probability of nulls coinciding in equal signal areas. The offset frequency determines the rate at which the combined instantaneous signal varies.

In order to achieve maximum intelligibility using a quasi-sync system, it is important that the signals arrive at the mobile receiver in phase and with a flat frequency response from approximately 300-2500 Hz. Frequency response and phase should also be consistent for any filters and land or radio links used in the overall system.

Multiple antennae for reception – antenna diversity

It was shown in Chapter 2 that signals can travel to a receiver by multiple paths and that the signal received is a vector addition of many signals which can vary in strength and phase. An improvement in overall reception is made by using more than one antenna for reception. The signals at any particular point in space vary depending upon the distance travelled and at any one point the signals can be good while at another, only a very short distance away, the received signals at an antenna can cancel. The situation is constantly changing with a moving mobile. An improvement in reception requires the antennae to be combined or switched depending upon the received signals. This adds to system complexity but can prevent the severe fading of signals.

The combining of antennae is more usual at base stations but it can also be incorporated, if required, in all mobile equipment.

Multicoupling of base stations

The effect on the environment of multiple sites for aerials around the country can be alleviated by combining the outputs of several base stations and feeding them into a common aerial. This also has considerable impact on the infrastructure cost of the installations. A typical installation is shown in Figure 11.3. In this system four base stations share a common omni-directional antenna for both receiving and transmitting. This is a broadband high gain antenna which is coupled to the receivers and transmitters by a low loss feeder and two bandpass filters.

Each transmitter is fed through a high Q cavity resonator with a typical Q of 11 000. This provides an isolation of 20 dB with a 150 kHz spacing. An additional 10 dB isolation can be attained by using series ferrite isolators and these also provide protection against the generation of intermodulation products in the power amplifier stage of the transmitter. The coupling to the transmitter SD (Spectrum Dividing) filter is made by a critical length of harness which provides the correct matching to the transmitter.

Receivers are fed from a high gain, low noise receiver splitter amplifier via the SD filter and a passive hybrid coupler network. The amplifier gain can be adjusted to compensate for the number of receiver ports used.

The advised maximum number of transmitters which can be coupled is 15. The minimum frequency spacing at low band VHF is 112.5 kHz, at high band VHF it is 100 kHz, 150 kHz at Band III (see later) and 200 kHz at UHF. Receiver splitter amplifiers are supplied with outputs of either 4, 8 or 16. When TALKTHROUGH mode is used it is advisable to use ferrite isolators.

Trunking

Computers and microprocessors have made possible control systems for mobile systems which allow them to operate automatically and have provided a range of facilities similar to those available on the PSTN digital exchanges. These are

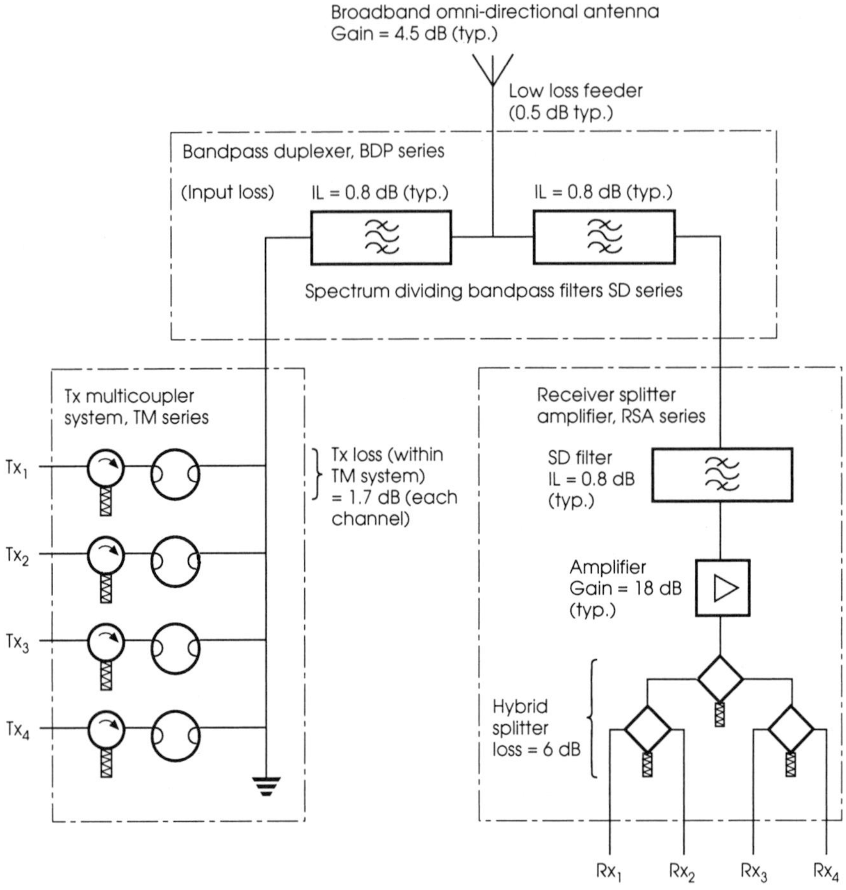

Broadband omni-directional antenna
Gain = 4.5 dB (typ.)

Low loss feeder
(0.5 dB typ.)

Bandpass duplexer, BDP series

(Input loss)   IL = 0.8 dB (typ.)   IL = 0.8 dB (typ.)

Spectrum dividing bandpass filters SD series

Tx multicoupler
system, TM series

Tx₁

Tx loss (within
TM system)
= 1.7 dB (each
channel)

Tx₂

Tx₃

Tx₄

Receiver splitter
amplifier, RSA series

SD filter
IL = 0.8 dB
(typ.)

Amplifier
Gain = 18 dB
(typ.)

Hybrid
splitter
loss = 6 dB

Rx₁   Rx₂   Rx₃   Rx₄

Figure 11.3 *Typical four-way multicoupler system. (Courtesy Aerial Facilities Ltd, part of the Aerial Group.)*

only possible with digital control. In addition PMR users who have been allocated a number of channels can co-operate with others and use a system called trunking whereby all the allocated channels are held in a common pool and allocated as required by a computer. This can improve the grade of service and maximise the use of the channels. Call queuing is a feature of these systems and if a channel is not immediately available the call is held until it can be allocated. Normally a time limit is allowed for each call after which it is terminated.

Control and signalling in such systems is performed on a separate channel to the speech channel. In conventional PMR a user is allocated a specific channel but, due to the shortage of available spectrum, this is invariably shared with another user and there is no privacy. In a trunked system, where the channels are pooled, the user occupies a private channel for the duration of the call. If a channel fails there is only a reduced grade of service and not a loss of communication.

In a large trunked network an area is arranged in cells and clusters and the available channel frequencies are divided amongst the cluster as in cellular radio (Chapter 9). The system requires a complex control system and signalling arrangement which is only possible with comprehensive computer control. Although more channels are available to each user the system requires the transceivers to scan all the available channels and this is only economically possible with low cost frequency synthesised oscillators and microprocessor control.

In order to establish initially a call between a base station and mobile there are basically two methods that can be used:

1. The receivers can monitor a dedicated control channel to listen for a call. If a radio unit wants to call, it communicates on the control channel. The control system allocates a specific channel for the duration of the call and informs all the relevant mobiles to tune to the channel frequency. When the call is complete the receiver returns to monitoring the control channel.
2. In an alternative method there is not a dedicated control channel and any channel can be designated. Receivers monitoring the system search for the marked control channel and when this channel is used for traffic the receivers search for the new control channel.

The size of the system determines which system is best. Large systems are faster and more efficient with method 1 while small systems with only a few channels use method 2. The two systems can also be combined with the control channel sometimes nominating itself for traffic and the radio units having the ability to scan for a control circuit. Such systems have the advantages of both 1 and 2.

The specification for the signalling protocol in the UK is MPT 1327.

Channel contention on the control channel

The problem with a single control channel is the possibility of two or more radio units trying to gain access to the channel at the same time. There are techniques available to avoid problems due to the clashes. These include:

1. Aloha (named after a technique developed at the University of Hawaii). This system provides no contention control.
2. Slotted Aloha (used on Band III systems – see later).
3. Carrier sensing multiple access (CSMA).

The Aloha system allows the mobiles to transmit irrespective of the state of the control channel.

The Slotted Aloha only allows messages to begin at the start of a fixed time slot. If messages clash, they coincide and only the particular time slot is lost.

CSMA does not allow a mobile to transmit when the channel is occupied. It is forced to wait for a random time and then try again. The time is arranged to be a random period in order to avoid continuous clashes.

National Trunked Network Band III

When a PMR user does not want the expense of establishing their own network they can subscribe to a public trunked network which offers both regional and national coverage. The frequencies for this network became available when the old monochrome 405 line television ceased to be used and part of the spectrum was reallocated to a trunked mobile network. The frequency spectrum is known as Band III and is between 174 and 216 MHz.

The service commenced in 1988 and since then the original system operators have combined their networks. An advantage to the subscriber is that the service is provided for a monthly subscription irrespective of usage and, therefore, the cost is known. Subscription costs depend upon the number of charging zones in which the subscriber requires coverage and whether communication between mobiles is required.

The system comprises a network of base stations throughout the country which are connected to regional traffic area switching centres (TASCs). Single or group calls can be set up and both data and speech can be transmitted. Calls are limited to 1 minute, after which time they are terminated, and, for commercial reasons, they cannot be connected to public telephone numbers. However, in the near future connections will be offered to a subscriber's PABX.

Data on the network can be sent to unattended vehicles and a security system can raise an alarm automatically in the event of a breakdown or tampering by sending pre-set codes (see Chapter 12 for other advantages). The provision of data in a mobile requires a modem at the sending computer and in the mobile and a laptop computer in the vehicle (Chapter 12). Successful tests have also been completed using bar code readers and miniature printers.

Communication on the trunked network

Communication on the system uses two frequencies but in a half duplex mode (cannot transmit and receive simultaneously − Chapter 7). There is an 8 MHz difference between the transmit and receive frequency with the base station to mobile being the lower frequency. Each channel has a 12.5 kHz spacing. The available spectrum at present for the Band III trunked network is 184.5-191.5 MHz and 192.5-199.5 MHz for mobile transmissions. Base station transmissions are 176.5-183.5 MHz and 200.5-207.5 MHz. The channels used by National Band III are 58-560. Only a quarter wave antenna with no gain is allowed to be used on the mobile and this is based on a centre frequency of 196 MHz. This is to avoid excessive power being radiated which can lead to the mobile retaining the control channel for too long. As the data travels further than the voice channel this can cause interference and lead to noisy calls and poor set-up. The high gain antennae, such as the 5/8 wavelength whip antenna, also have a narrower bandwidth and this can cause a high incidence of one-way calls.

The network comprises three traffic area switching centres (TASC) at Manchester, Wembley and Coventry. These are associated with the traffic area controller (TAC) and are situated together. Data communication to the base stations from these points is transmitted on 9.6 kbit serial data links.

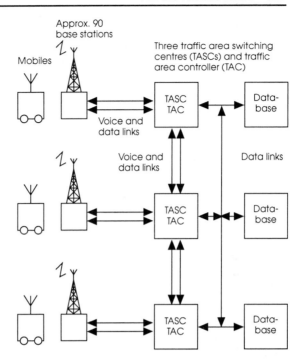

Figure 11.4 *National Band III network structure.*

The country is divided into eight regions and base stations are fed in such a way as to provide alternative paths to TASCs should a particular link fail. Thirty-two designated control channels are used throughout the country on which all signalling and validation of subscribers is performed.

Information regarding subscribers' particulars is kept in a database which can be anywhere on the system. This is the management network centre (MNC). In practice they are duplicated at each TASC and each can be brought into use as required. They, therefore, act as a back-up to each other should links fail. Each databank is also duplicated on each site. When a subscriber enters a region the ESN (Equipment Security Number) is registered and validation for use in the particular region is obtained. The location of the mobile is also recorded. The ESN is programmed into the radio set at the time of manufacture and cannot be changed. A typical number consists of:

Three digits for manufacturer
Three digits for model number
Six-digit serial number

In addition all units are given a network number of 10 digits arranged in three parts. This consists of a prefix, fleet number and unit or group number. This allows a shortened number to be dialled for regularly used numbers. Control codes are also used to obtain different facilities.

Base stations

There are approximately 90 base stations sited around the country. They are arranged around regional groups and linked to a particular TASC. The TASCs are themselves linked together with alternative routes should a link fail. The base stations differ in the number of channels transmitted and this is determined by the traffic intensity. 6, 12- and 24-channel base stations are installed. The transmitters are normally fixed frequency and are combined on the output to the antenna. A few frequency agile transmitters are available which can be used in times of high demand or can replace a failed unit. Base stations are designed to allow local communication should links to the network fail.

Traffic Area Switching Centre

This consists of three main sections:

1. It is a digital System X exchange which switches digital voice transmission between the base stations.
2. It is a packet switched network centre which transmits both the data generated by the user and operational data.
3. It is a digital computer which stores all the subscriber data and movements. It is used to assist a link-up across the network.

Operation of control channel

The Slotted Aloha arrangement is used on the control channel to improve contention conditions. Each time slot on the forward channel from the TSC to the mobile contains two codewords which give the system identity and the message. The first codeword is followed by a preamble and synchronising sequence with check bits. The mobile radio unit starts its transmitter and sends a preamble and synchronising sequence followed by a one-word message. This is timed to coincide with a forward channel slot. The type of messages sent include:

1. Requests from radio units for calls.
2. Acknowledgement by base station of requests.
3. Allocation of traffic channels.
4. Responses demanded by the TSC to an addressed mobile.
5. Acknowledgements from mobile radio units.
6. Aloha messages to the mobiles from the base station to allow and to control access to the system.

The system uses a 20-bit address comprising a prefix of 7 bits and a 13-bit identification. This provides a million unique codes ($2^7 = 128$, $2^{13} = 8192$, $2^7 \times 2^{13} = 1\,048\,576$).

A radio unit can have more than one identity as it can belong to groups. When calling a base station a radio unit must supply the following information:

1. The identity of the radio unit.
2. The identity of the individual or group the radio unit is calling.

3. Whether data or speech is required.

4. Whether the called radio units in a group are to talk back.

5. The degree of priority the call must have.

Radio units when requested must signify their ability to communicate and this is normally implemented by the microphone hook switch (ready for communication control – RFCC). No channel is allocated until a called subscriber is ready to accept the call. Information is passed back to the caller indicating whether the called subscriber is ready and whether the call is queued. If in a queue the caller can abandon the call. This, however, can be a disadvantage as the abandoned queue position is not filled due to the method of queuing. If a subscriber immediately recalls the mobile the queue is unnecessarily extended as the caller rejoins the end of the queue. This can lead to a 'System Full' message being received by the caller.

When the subscribers are ready a channel is allocated and both parties are instructed to 'Go to a Channel'. A signalling burst for 80 ms is required from the caller's transmitter every few seconds to indicate it is still on. The audio is muted during this period and the signal is not heard. A signal is also sent each time the press-to-talk (PTT) button is released. A call is disconnected by the first subscriber to replace the phone on hook unless it is a group call. In this situation the call is only disconnected by the caller. Failure to disconnect by the subscriber leads to a disconnection by the base station if a transmission is not detected over a period of 7 seconds.

Facilities exist for an indication in the mobile or at the dispatcher if, during their absence, a caller has tried to contact them. Some mobiles and dispatchers also have a call logging facility whereby a mobile or dispatcher can be in a call logging mode and all units calling the mobile or dispatcher are logged and stored in memory.

In addition to data being sent over the network it is also possible to send status messages. The number of different status messages is determined by the mobile type and programming is required. In some instances a simple status number is displayed. On other equipment up to 30 different messages can be displayed in alpha-numeric form on a display in the mobile. These messages are fixed and must be programmed for a particular fleet who may use the facility.

The efficient performance of a trunked service depends upon the control channel not becoming overloaded. If it does, the grade of service is drastically reduced. Many factors, determined by both the make-up of the code and its performance in an operational environment, including multipath propagation and moving mobiles, affect the efficiency of a trunked network. The MPT 13271327 protocol is based on the Aloha system and is designed to:

1. Ensure access delays are a minimum.

2. Ensure the system is stable.

3. Achieve maximum traffic even in busy hours.

4. Control the effects of different mobiles trying to access the system simultaneously.

The trunked networks offer considerable advantages to many users and provide an efficient use of the available spectrum. There has been considerable growth in the USA since their introduction in 1979 and the same pattern is being repeated in the UK.

12 Mobile data

Computers and faxes have become essential tools to businesses. For the millions of people who travel as part of their work the ability to access information held on office computers or to update records held on central computers with information from the field can save time and money and provide efficiencies which were not possible only a few years ago. Engineers, sales staff, service technicians, managers and all the professions can benefit from a readily available databank which can be used on an interactive basis. Orders can be placed, deliveries arranged and invoices prepared directly from the field while service engineers can list the parts used and time taken directly from their place of work. Long haul lorry drivers can send and receive documents from their home base which may avoid delays at borders and deliveries to customers. It is also possible for the transport manager to control the fleet more efficiently. The written message is usually both more cost effective and accurate than the equivalent spoken message. It also provides a permanent written record. The digital data travels much quicker than speech and is limited to the necessary information that is required. Most spoken messages contain social conversation not applicable to the particular work in progress. Messages can be repeated until received without errors and this has proven useful to drivers in heavy traffic who need to receive instructions but whose concentration is required on actual driving at the particular moment a message is received. Taxi drivers and couriers are now using data terminals in their vehicles in order to receive instructions about the next fare pickup or delivery.

Digital signals

The data is produced as binary digits. These are pulses whose voltage amplitude determines whether they represent a 1 or 0 (Chapter 7).

Groups of pulses are arranged to form a code and how many pulses are used determines how many individual pieces of information can be represented. The Baudot code used for telex transmission uses 5 bits and, therefore, $2^5 = 32$ pieces of information can be represented by arranging the 5 bits in all the possible permutations. This can be increased by using one of the codes to indicate letters are following and another to indicate numbers are following. The same codes can then be reused.

Although there are sufficient individual codes for all the letters there are insufficient to allow upper and lower cases to be used. The code which allows this is the American Standard Code for Information Interchange (ASCII), the normal code used for digital communications. It uses 7 bits for each character. The first 3 bits determine whether a letter (upper or lower case), number or character is transmitted and the last 4 bits determine the actual letter, number or symbol.

A pulse contains an infinite number of odd harmonics and if these are not all transmitted in the correct phase relationship to each other the pulse is distorted. The DC content must also be maintained otherwise the received signal does not reach its correct voltage. A digital 1 is only recognised as a 1 if its voltage has sufficient amplitude otherwise it is considered to be a digital 0. Audio amplifiers and transmission bridges used in analogue circuits remove the DC. The audio analogue channels also restrict the bandwidth to 300–3400 Hz on a normal telephone channel (commercial speech) and, therefore, it is not possible to transmit digital signals on an analogue circuit for any considerable distance. As shown in Chapter 7 the frequency of a digital signal is dependent upon the pattern of ones and zeros and a word consisting of either all ones or zeros produces zero frequency which could not be passed.

In order to transmit the data over an analogue network or radio link it is necessary to convert the signal into a form which is acceptable to the network. It must, therefore, be modulated so that it has the characteristics of an analogue signal and can be transmitted without distortion. At the receive end the signal must be restored to a digital format by a demodulator. The equipment that combines the modulation and demodulation processes at a terminal is known as a modem. The type of

(a) Original pulse voltages

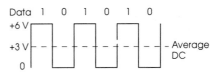

(b) If DC content is lost the pulse voltages range between +3 and –3 (note pulse is still 6 V but peak voltage now only +3 V)

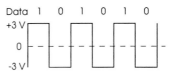

(c) If DC content is lost pulses become 0 V. Average DC content of a data waveform varies with the bit pattern

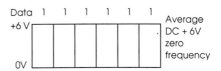

Figure 12.1 *Variation of peak voltage with loss of waveform DC content.*

modem that is required depends upon the modulation process used for the transmission and this depends upon the number of bits (digital pulses) that need to be transmitted each second.

Types of modulation for modems for analogue circuits

Three possible forms of modulation can be used:

1. Amplitude modulation (AM)
2. Frequency modulation (FM)
3. Phase modulation (PM)

In defining the system the term shift keying is added after the first letter to identify the use of the modulation process for digital transmission (e.g. FSK – Frequency Shift Keying).

When deciding which modulation system to use consideration must be given to:

1. Data rate of digital signal.
2. Complexity of equipment and cost.

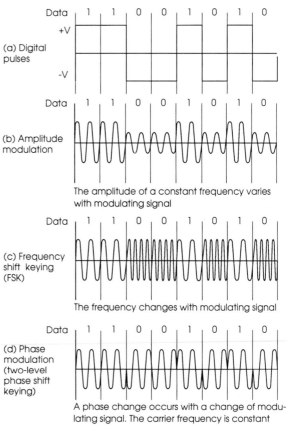

Figure 12.2 *Two types of phase modulation are used:*
  1. Differential phase modulation in which a change of phase simply indicates a change from 1 to 0 or 0 to 1. No reference signal is required to detect the signal.
  2. Fixed reference phase modulation in which a 1 is indicated by a continuous signal whose phase is referenced in one direction and a 0 is indicated by a phase inversion of the signal. Demodulations requires the detection of the phase relationships and a reference is required

(a) Digital pulses

(b) Amplitude modulation

The amplitude of a constant frequency varies with modulating signal

(c) Frequency shift keying (FSK)

The frequency changes with modulating signal

(d) Phase modulation (two-level phase shift keying)

A phase change occurs with a change of modulating signal. The carrier frequency is constant

3. Bandwidth required for transmission of data.

4. Performance of modulated signal in the presence of noise and multipath propagation.

All the systems can be modulated by either two-level or multi-level techniques. In the simplest form the carrier is modulated by either a digital 1 or 0 and the carrier has two states. However, digits can be combined in pairs so that the four states of digital information relate to the possible adjacent combinations 00, 01, 10, 11. The system can be continued using tribits producing eight possible combinations. Each combination reduces the apparent bit rate and, therefore, the modulating frequency and bandwidth requirements. This is, however, at the expense of more complexity in the modulation and demodulation process.

When detection takes place at the receiver the system can function in a coherent or non-coherent mode. In the coherent mode a reference signal in the form of a pilot tone or one obtained from the modulated signal must be present. This must be closely matched to the phase and frequency of the received carrier. A non-coherent system does not require the phase reference and this system is common in FSK. Coherent demodulation can be used for FSK which produces superior performance in the presence of noise but it requires two oscillators in the detector which have the same frequencies as the modulating frequencies and this introduces expense and complexity. After the detection there is usually some circuitry that examines the detected signal and makes a decision about the correct data waveform.

Amplitude modulation (Figure 12.3)

In an amplitude modulation system a 1 or 0 is transmitted as two different amplitudes of the same carrier frequency. In practice, during transmission, noise and non-linearity caused by the system components can both distort the signal levels and lead to a wrong data signal being demodulated. Amplitude modulation is used as part of other systems such as quadrature amplitude modulation (QAM).

Frequency modulation – frequency shift keying (FSK)

This system has been referred to in previous chapters and is the common format for coding data with bit rates from 300 to 1200 baud (pulses per second). The system uses two frequencies to represent a 1 and 0. It was shown in Chapter 3 that the frequency modulated wave consists of the carrier and sidebands which appear either side of the carrier. The information is carried in the sidebands and the importance of the individual sidebands depends upon the power in the sideband. If the important sidebands of the frequency modulated waveform are to be contained in the audio channel bandwidth the signal must be limited to the first-order sidebands.

It was shown in Chapter 7 that the fundamental frequency of a digital waveform is half the bit rate. The 1200 bits/s data, therefore, has a fundamental frequency of 600 Hz (1200/2) and the bandwidth required to transmit the first-order sidebands is 1200 Hz:

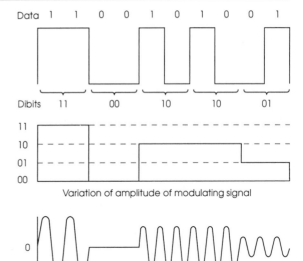

Variation of amplitude of modulating signal

Figure 12.3 *Multi-level amplitude modulation.*

Multi-level amplitude modulated signal

Bandwidth = bit rate (for first-order sidebands = 2 × modulating frequency). This is within the audio bandwidth. When transmitting 1200 bits/s on the PSTN system, the 1 is transmitted as 1300 Hz and 0 is transmitted as 2100 Hz. The nominal carrier frequency is the arithmetic mean and is 1700 Hz. The deviation is, therefore, 400 Hz and must be a compromise between providing a good signal-to-noise ratio and a low modulation index to produce only low order sidebands (Chapter 3).

When using two frequencies for FSK to represent a 1 or a 0 the bandwidth required for a data rate above 1200 bits/s (e.g. 2400 bits/s) exceeds the available bandwidth of commercial speech. It is then necessary to use multi-level modulation and reduce the data rate by combining bits to form dibits or other combinations (see RD-LAP format). A greater number of carrier frequencies are required to carry the different modulating signals. Phase modulation is commonly used for data rates exceeding 1200 bits/s (see later).

Fast frequency shift keying (FFSK)

This system is used for signalling in Band III (see Band III signalling code) where the data is transmitted at 1200 bits/s. In order to simplify decoding the bit transitions

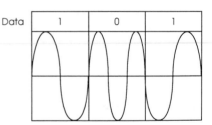

Figure 12.4 *Fast frequency shift keying (FFSK): 1200 Hz is transmitted for 1 (1 cycle); 1800 Hz is transmitted for 0 (1 ¹/₂ cycles). There is no phase change between 1 and 0.*

always occur at the zero crossover points which, provides a waveform without changes in phase. This is achieved by transmitting one cycle of 1200 Hz to represent 1 and one and a half cycles of 1800 Hz to represent 0.

Phase modulation

The practical limit on a telephone line or speech channel restricted to a commercial speech bandwidth (300–3400 Hz) is 1200 baud. When every bit is encoded as a single cycle the bit rate and baud rate are equal (1200 bits/s). To increase the bit rate on a system the transmitted bit rate must be modified to appear to be a lower bit rate and, therefore, a lower modulating frequency. Phase modulation is extensively used in several forms to accomplish the higher data rates.

Two-level phase shift keying (Figure 12.2)
In its simplest form phase modulation changes the phase of a carrier by 180 degrees depending upon whether a binary 1 or 0 is transmitted. To demodulate the signal the waveform is added to the reference carrier at the receiver and the signals either add or subtract and produce the original space or mark (0 or 1). This system obviously does not reduce the bit rate.

Quadrature phase shift keying (Figure 12.5)
Instead of using a single carrier it is possible to split the single frequency carrier into two separate carriers by using a 90 degree phase shift network for one signal. This provides two independent carriers of the same frequency which can both be modulated (same system used for colour TV transmission). Each carrier can be modulated by a phase change of 180 degrees and, therefore, four signals can be modulated. The four modulated signals now have a phase relationship to each other of 0, 90, 180 and 270 degrees. The modulating signals are dibits (see below).

In order to demodulate the quadrature phase encoded signal the incoming signal is added to the reference waveform and the signal is obtained from both the polarity and magnitude of the resulting waveform.

Dibits and multi-level coding
When bit rates exceeding the available bandwidth of the communication system need to be transmitted their apparent rate is reduced.

In practice multi-level coding is used whereby adjacent digits are paired and used as a single modulating signal. Four different dibits are the possible permutations of adjacent digits and these are 00, 11, 01, 10.

The use of the pairs or dibits reduces the bandwidth requirements by reducing the fundamental frequency. If dibits are used the fundamental frequency is reduced by half. It, therefore, becomes 1/4 × bit rate. Phase modulation and frequency modulation (also known as angle modulation) require the same bandwidth when modulated with the same fundamental frequency as they produce the same type of sidebands. By using dibits a data rate of 2400 bits/s can be transmitted on a

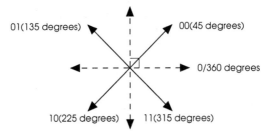

(a) CCITT specify 45 degrees for 00 dibit to assist recovery of synchronisation signal when a long-stream of zeros occur. Older systems used 0, 90, 180 and 270 degrees.

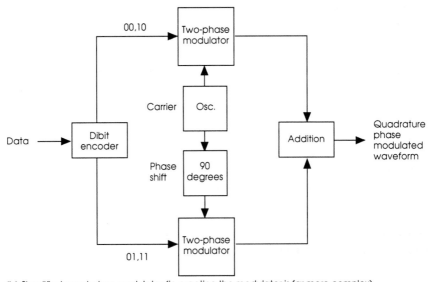

(b) Simplified quadrature modulator (in practice the modulator is far more complex).

Figure 12.5   *Quadrature phase modulated signal.*

commercial speech audio channel. The idea of reducing the fundamental frequency can be continued by using tribits and eight phases. This allows a higher data rate to be coded and transmitted within the speech channel but the modulation systems become complex. Tribits using 8 phases can transmit 3600 bits/s on a 1200 baud transmission system and using 16 phases for groups of 4 bits allows 4800 bits/s.

When using dibits or higher groups, the transmitting and receiving equipment must be synchronised in order to identify correctly the groups of bits. The synchronisation signal is derived from the changes of phase in the received signal.

Demodulation of quadrature phase encoded signals is achieved by adding the reference waveform to the signal but both the polarity and magnitude of the sum are used to decode the dibits.

Differentially encoded phase shift keying (DPSK)

The previous phase systems require the use of a reference signal at the receiver which must be an accurate copy of the transmitting reference signal. This must be accurate in both amplitude and phase. In some phase systems this signal is derived from a pilot tone sent together with the encoded signal. However, this requires additional bandwidth which may not be available.

To overcome this problem a system known as differential phase shift keying is used whereby the use of a reference is not required. This is a method whereby no particular value is attributed to the phases. It is the change in phase from one cycle to another that indicates the data bit value. In a system using one phase for 1 and a 180 degree phase change for 0, the phase of the carrier is altered by 180 degrees each time the leading edge of a 0 occurs but it is unaltered by a 1. In the detection stage a system can be used which does not require coherent detection. The data is recovered by the current bit being multiplied with the previous bit from the output of a 1-bit delay line. The demodulation becomes the reverse of the modulation process, a 180-phase change becomes a 1 and no change is interpreted as 0.

DPSK has advantages of simplicity over systems requiring the carrier recovery which can also be degraded by phase jitter due to noise. These systems also have the possibility of locking up 180 degrees out of phase. However, the performance of simple demodulated DPSK can be inferior to coherently demodulated phase shift keying as it is multiplied by a delayed one-line bit, which itself can be noisy, rather than recovery being achieved with a clean carrier. As 2 bits are required to decide a 1 or 0, errors tend to occur in pairs. More complex demodulation systems using coherent detection can be used with improvements (DCPSK – Differentially Coherent Phase Shift Keying).

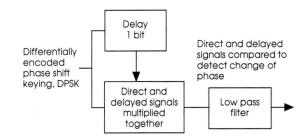

Figure 12.6 *Simple demodulator for DPSK signals.*

Quadrature amplitude modulation (QAM)

This modulation system combines phase modulation with amplitude modulation. The method uses a constant carrier frequency which is arranged in phases. Each phase is amplitude modulated by a combination of data bits arranged in groups as previously explained. CCITT V22 bis (or b – bis means second revision) is the specification for 2400 bits/s using QAM in full duplex with a 600 baud rate. V32 specifies 12 phases with 3 amplitudes (see diagram). Sixteen combinations are used

Figure shows 12 phases and three amplitudes. Modulation signals are quadbits (combinations of four adjacent digits). These provide 16 combinations. Used for transmission of 4800 bits/s at 600 baud.

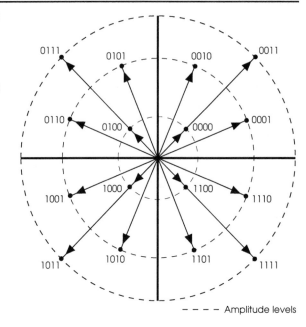

Figure 12.7 *QAM – Quadrature amplitude modulation.*

– – – – Amplitude levels

for 4800 bits/s. Quadbits (combinations of four digits) or nibbles are used to modulate the carriers.

A number often appears before the letters QAM and this indicates the number of amplitude levels on each phase (e.g. 4-QAM means quadrature phase modulation with four levels of amplitude modulation).

Similar numbers appear before PSK and FSK and these indicate the number of phases and frequencies used respectively.

Summary

There are many variations of the different modulation systems that can be used for digital coding. These variations affect the bandwidth and power required and also determine the sensitivity of the data to distortion. The apparent data rate that ultimately determines the bandwidth requirement can be reduced by combining the adjacent bits into groups of various lengths before modulation.

Amplitude modulation

All the methods of amplitude modulation given in Chapter 3 are possible but if double sideband is used it can waste power and bandwidth. Vestigial sideband transmission (Chapter 3) is used to reduce bandwidth requirements when transferring high speed data on transmission systems which directly connect two computers. Amplitude modulation is used together with phase modulation to produce QAM.

Frequency modulation

In the simple two-level systems frequency modulation is used for data rates up to 1200 bits/s. In the higher levels (three, four, eight) it is used to adapt FM analogue radio systems to digital transmission. Coherent or non-coherent detection systems can be used. Coherent systems offer advantages where noise levels are high at the expense of complexity and cost.

Phase modulation

Two-level phase modulation can be used with either coherent or differentially coherent detection and the systems are tolerant to distortion but waste bandwidth. QPSK using four levels of modulation can be detected with coherent or differentially coherent methods but coherent is preferred where interference may occur on high capacity digital radio links. An increase in levels makes the system more susceptible to distortion but an eight-level system with coherent detection provides good bandwidth efficiency and can be used with other analogue channels.

For the reader who wishes to study these systems further there are many books on digital coding but they tend to be very mathematical and are beyond the level intended for this book.

Data on the cellular network

The cellular system is required to transmit the data as an analogue signal within the bandwidth of a channel. In order to do this the data rate must be reduced if it exceeds 1200 bits/s and the data must be modulated to appear as an analogue signal. This is performed by a modem.

The transmission of data over telephone circuits is governed by a series of recommendations from the CCITT known as the V series. An additional X series governs the attachment of terminals to the public data network and provides a set of standards for interfacing terminals to networks.

Cellular radio inherently has a poorer signal-to-noise ratio than a normal line and consequently is liable to short breaks in transmission. Although not noticeable during a conversation, the break can be sufficient to interrupt data and cause an error. The modems must, therefore, incorporate error correction in order to provide a satisfactory transmission. The correction needs to be incorporated in both the transmit and receive equipment for the best results. Modems incorporating error correction are designated V42 or MNP4.

The V22 standard is for the transmission of 1200 bits/s and in conjunction with V42/MNP4 provides the most reliable transmission of data over the cellular system. V22 bis is the standard for 2400 bits/s full duplex but is not recommended on the cellular system. For higher data rates a standard known as V32/4800 provides 4800 bits/s and is acceptable on the network.

Higher data rates require data compression techniques whereby coding the data can produce an apparent lower data rate for transmission. Using the V42 bis standard a data rate equivalent to 2.5 times the transmission rate can normally be achieved.

The coding used for V22 1200 is DPSK (Dual Phase Shift Keying) and for V32/4800 is QAM (Quadrature Amplitude Modulation – see previous section).

The full specifications for all the V series can be found in the International Telecommunications Union, CCITT, Blue Book. This is called Data Communication Over the Telephone Network.

Error correction and compression

A major advantage of digital communication is that within the conversion error correction and data compression can be built into the system. As previously seen with GSM and paging, error correction is obtained by splitting the data into blocks and adding checksum information in the form of additional bits. At the receiver the additional information is used to determine whether the data arrived without error and possible corrections can be made by the use of Hamming codes. A compromise has to be reached in the size of data package. A large packet allows maximum throughput on the system but under noisy conditions is more likely to be corrupted than a smaller packet.

Data compression methods strip the incoming data to the modulating system of all redundant data prior to transmission. This may take the form of transmitting either most significant bits or least significant bits depending upon whether a large amount of the data is the same. A further technique is to examine the frequency with which certain characters occur and assign a reduced length character for the most common ones. Run-length encoding is also used whereby a string of similar characters are compressed to a single character and an additional character is added to indicate the length of the string.

Modems can also include a serial data buffer that allows a modem designed for 2400 bits/s operation to accept 9600 bits/s on the input. Such data is compressed and sent to line as the conditions allow. The transmission terminal must be held up until the buffer can be cleared if noise prevents transmission and reception from the modem.

Modems with a buffer can be used to work automatically to adjust to the speed of the receiving modem. This is achieved by storing the data in the buffer and sending it as required.

Radio Data Link Access Protocol (RD-LAP)

This protocol has been developed to provide high speed radio communication for a dedicated radio network using packet switching. It allows rates up to 19.2 kbits/s to be used, which is the highest rate presently available, but it has the capability of higher rates as systems evolve. It allows unrestricted roaming within the network and provides a high channel efficiency by using a slotted digital multiple access system (slotted DSMA – see later explanation). This system is used as the air interface on the Motorola designed digital data networks.

The RD-LAP specification is fundamentally a link layer protocol in order to transfer data from one end of a link to another. To perform this it details aspects of the lower three layers of the OSI reference model (Chapter 7). These layers are as follows.

1. Physical layer.

Within this layer the RD-LAP specifies the communication process between the mobile and the network infrastructure. This includes the method of baseband signalling, carrier modulation and RF channel structure. It also specifies the electrical and mechanical interfaces.

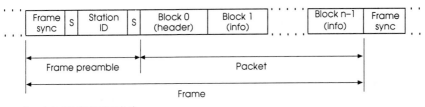

S = channel status symbol

RD-LAP frame structure

(a) Each packet contains user data and end-to-end control information.

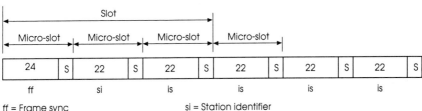

ff = Frame sync
S = Channel status symbol

si = Station identifier
is = Information symbols

RD-LAP outbound channel slotting

(b) A channel status symbol is interleaved throughout the frame to control inbound channel access. Mobile terminals can only attempt inbound transmissions when the CSS is in the idle state. A packet fills more than one slot and continues to be transmitted until complete.

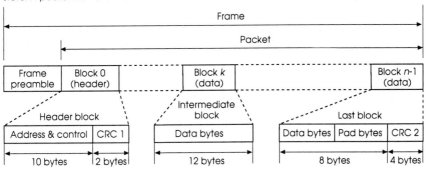

RD-LAP packet format

(c) CRC = Cyclic Redundancy Checking.

Figure 12.8 *RD-LAP format (Diagrams courtesy of Motorola.)*

2. Data link layer

In this layer the protocol specifies the RD-LAP frame format, channel access and method of correction and detection.

3. Network layer

In this layer the RD-LAP specifies how data is transferred from one end of the link to the other.

Radio channel

The radio transmission is performed over a full duplex circuit using narrow band FM radio channels between 400 and 900 MHz. The specification allows either 12.5 or 25 kHz channel separation. The base sites operate in full duplex while the mobile terminals operate in half duplex mode. Data is transmitted as a frequency modulated carrier using four-level Gaussian frequency shift keying (FSK). The baseband baud rate can be either 9600 bits/s or 19.2 kbits/s for 12.5 kHz and 25 kHz channel spacing respectively.

The RD-LAP specification specifies transmitter and receiver operating standards for output power, stability, deviation, emissions and distortion.

RD-LAP format (Figure 12.8)

With the exception of frame synchronisation, all data, addresses and control characters are produced as 8-bit octets (bytes). The format is shown in Figure 12.8. Error detection codes are used both at the frame level and the packet level.

The data is serialised and broken into tribits (eight permutations of adjacent bits). Each tribit is converted into two, four-level symbols which are interleaved (see GSM for principle) before transmission using four-level FSK modulation across the radio channel. A channel status symbol (CSS) is inserted every 22 symbols throughout the frame and this is used to control the inbound channel access.

Slotted DSMA

All outbound transmissions from base stations to mobiles are arranged on a first-in-first-out basis by the network system but inbound traffic is controlled by the CSS in order to avoid collisions. The CSS is set either to Idle or to Busy by the base station depending upon whether it senses a mobile has accessed the inbound channel. Inbound traffic is also timed to occur in specific time slots and this increases the channel efficiency. Figure 12.8(b) shows outbound channel slotting.

Packet format

The format consists of a header block followed by 43 data blocks each containing 12 bytes of data. Padding may be used in the final block to ensure there are 12 bytes. The link management depends upon the header as it contains information relating to:

1. Whether the direction of the message is inbound or outbound.
2. The packet format.

3. The class, type and status of the packet.
4. The outlet service points to which the data must be directed.
5. The number of data blocks in the packet.
6. The sequence numbering of the packets.
7. The amount of data padding that has been used.
8. How much data header exists.
9. The cyclic redundancy checking code for the block header.
10. Whether a received message confirmation is required.

Sequencing of packets.

Both the sender and receiver of the packets have a pair of registers that are initialised via a flag in the header block. The registers specify the sequence number of a received or transmitted message. When transmitted the order number is inserted in the header block of the outgoing data. When it is received and decoded the number is checked against the number the receiver expected and if correct it is acknowledged back to the transmission end. If the number is one less than expected the receiver realises it is a repeat message. Other mismatched numbers cause the message to be abandoned and an out-of-sequence message to be sent back to the sender.

The versatility of the RD-LAP specification allows considerable management control of the network and the possibility of providing connections and services. A full study of the specification is beyond the level of this book.

Channel capacity of a digital system

The data rate that can be reliably used on a digital circuit is governed by its bandwidth and signal-to-noise ratio. The maximum data rate can be calculated from the Shannon-Hartley theorem. This states that:

$$C = B \log_2 (1 + S/N)$$
where C = channel capacity in bits/s
B = bandwidth in hertz
S/N = signal-to-noise ratio in absolute terms (not decibels)

The formula takes into consideration only white noise. It does not consider impulsive noise.

Example
The absolute S/N of an audio channel is 512 and it has a bandwidth of 3 kHz. What data rate can be transmitted?

Substituting in the formula:

$C = 3000 \log_2 (1 + 512)$

It should be noted that the logs are to the base 2 not 10:

$C = 3000 \log_2 (513)$

To find $\log_2 513$ we use the standard form of logs:

$$\text{Number} = \text{Base}^{\log}$$
$$513 = 2^n \text{ (where } n = \log \text{ of the number)}$$
$$9 = n$$
$$C = 3000 \times 9 = 27\,000 \text{ bits/s}$$

This is the theoretical limit for the channel and exceeding this data rate results in very high errors. In practice the rate used is lower in order to improve the reliability.

Mobile data networks

A subscriber wishing to use data over the public mobile radio links has the options of using the existing cellular or Band III systems, which also carry speech, or one of the dedicated data networks. The cellular and Band III networks are analogue networks and data, therefore, must be modulated to a form that allows it to be transmitted as an analogue signal. As previously described the modulation forms depend upon the data speed and the bandwidth available. A dedicated data network, however, can operate as a digital system with all the advantages that this can provide. Subscribers must, therefore, decide on all the facilities they require from the network before deciding which network is most suitable for their requirements.

A dedicated data network, such as that designed by Motorola, uses a packet switched data network (see Chapter 7 for packet switching). Packet switching networks operate by the customer's data being arranged into small packets of information, complete with their source and destination address, and being interleaved on the network with other users. It is a very fast and cost effective method of sending data and allows two-way interactive access to databases, files and software. The system requires no call set-up or disconnection procedures.

The switched networks maintain an open line for the duration of the call irrespective of whether data is being transferred and consequently the systems are more expensive and possibly more suitable for the casual use of data.

Dedicated mobile data network

The Motorola designed data system operates with a high speed radio protocol (RD-LAP) which operates at 9.6 kbits/s. As it is digital, sophisticated error correction systems can be incorporated as previously seen on other digital systems.

The system operates at frequencies around 450 MHz and consists of 20 channels. A mobile can roam freely throughout the network without the need for the subscriber to change channels or have any knowledge about the system. The modem automatically transmits a 'Here I Am' message immediately it moves out of the range of one base station and into the area of another. The network, itself, is designed on a modular basis to simplify expansion whenever traffic demands additional facilities. An open network approach has been adopted allowing interfaces to many different systems. This allows a wide range of applications which can be distributed throughout the network.

The Motorola data network comprises five main components. These are:

1. Terminals and radio modems
2. Base stations
3. Backbone network
4. Operation centres
5. Network management

The system has been designed to allow any terminal with an RS232 asynchronous serial port to be connected to a radio packet modem which provides an input to the system. Asynchronous operation is a method of data transmission in which each character is framed by bits for start and stop signals. The start bit triggers a timing mechanism in the receiving terminal which proceeds to count the succeeding bits of the character as fixed time intervals. The receiver is reset by the stop bit and becomes ready to receive the next character. This method is used for slow speed data rates. High speed data rates use a synchronous system whereby synchronisation is maintained by clocks in the transmitting and receiving terminals. Data is transmitted without gaps but should gaps occur in the data they are filled with Idle bits. Synchronisation is obtained at the beginning of a transmission by special characters which align the transmitting and receiving clocks.

The national network consists of over 100 base stations which are situated on buildings and hills and each base station covers a specific area of the country. The radio modems communicate with the area base station using the RD-LAP protocol. Each base station is connected to the backbone system through dedicated lines operating at 9.6 kbits/s.

The backbone network is a digital frame relay network where data speeds of 2 Mbits/s are used. The network uses eight local concentrator points located in different parts of the country and each concentrator has alternative routings to the operation centres. In all digital networks the transmission data rates change throughout the system. Information starts from a single source and it is gradually concentrated with other sources for transmission on the network. As the traffic increases on a particular route the data rates are increased in order to carry all the traffic. This requires circuits which are capable of the higher transmission rates. One advantage of digital transmission is that information can be placed in a store at one rate and read out at a different rate. Using this process the data rate can be either increased or decreased.

There are two independent operational centres which provide back-up for the network management should a problem develop at one site. The operational centres control the distribution via an Ethernet LAN network to the switching nodes. The switching nodes are the area communications controllers and are situated within the operational centres. The area communication controllers are responsible for routing the data, managing the network, controlling the base sites and converting the data to and from the internal radio protocols. They also provide authorisation and validation of the subscribers and billing information. In addition links are provided to the customer computer hosts. Data is received as packages at the data switches where

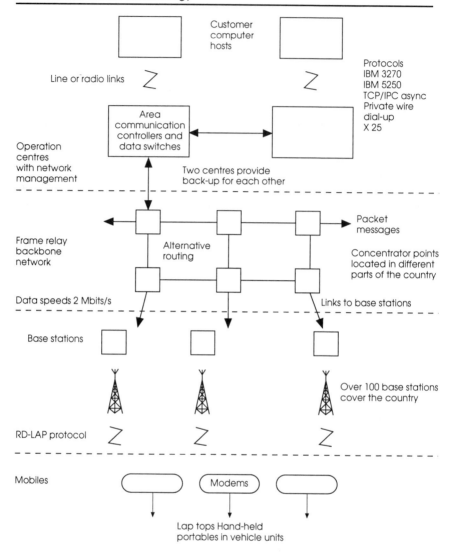

Figure 12.9   *Dedicated digital data network.*

they are converted and routed to the customer in the appropriate protocol such as X25 (CCITT recommendation for the protocol for packet switching networks) and IBM protocols etc.

The network management is performed in the LAN network situated within the operational centres where information is continuously gathered about the performance and possible faults on the system.

Customer terminals and modems

The interface between the customer's mobile equipment and the radio network is through a radio packet modem. This is compatible with a wide range of manufacturers' portable terminals. There is also a vehicle mounted modem which fits into the dashboard or boot and this can be accessed through a portable computer or a fixed terminal. The modems are frequency agile which allows them to tune to any of the available channel frequencies. The vehicle mounted modem allows fast full duplex operation at 9.6 kbits/s and connects with a standard cable to any PC or hand-held terminal that has a standard RS232 port. It is powered from the vehicle's electrical system.

Portable models can be used with batteries and these have either a built-in antenna or an extendible one. Manufacturers will also produce portable computer equipment which includes the internal modem and this will become more apparent in future years.

There are other designs of dedicated data systems operating in the UK with different operators.

13 Cordless telephone

In the 1980s the cordless telephone began to appear in the UK. Initially these came from abroad and were unauthorised for connection to the telephone system. However, as other models began to be approved their popularity increased due to the freedom of movement the phone offered around the house and the garden.

The cordless telephone consists of a base unit, with a transmitter and receiver, and a portable phone, again with a transmitter and receiver and the mechanism for dialling. The base unit operates from the mains power and the telephone has a battery. The base unit also contains a charger for recharging the telephone battery.

Communication between the phone and base station is by a number of fixed channels. Two frequencies are used for full duplex working. Because the telephone uses a standard analogue modulation system and a limited number of frequencies there is the possibility of interference from other users on the same channel and also the system can be noisy. There is no security for the conversation and anyone on the same channel and within a short distance from a user can listen to the call. The distance travelled by the radio waves is, however, limited to about 75 metres and in practice is often less due to the environment and the positioning of the antennae.

Cordless telephones use one of eight different pairs of frequencies. These are factory set although some give the user the possibility of changing channels if they are experiencing interference. The frequencies used are given in Table 13.1.

There is a special code used which allows the base unit to work only with the supplied handset. This avoids the possibility of a handset on the same channel routing phone calls and bills through another person's line. It is also possible when another person on the same channel is operating in close proximity to another person's base station that they can hold a call after it has been cleared down. In these circumstances a user must ensure their call is cleared.

Table 13.1

Channel letter	Base frequency (MHz)	Handset frequency (MHz)	Channel no.
A	1.642	47.45625	1
B	1.662	47.68575	2
C	1.682	47.68125	3
D	1.702	47.49375	4
E	1.722	47.50625	5
F	1.742	47.51875	6
G	1.762	47.53125	7
H	1.782	47.54375	8

Digital cordless telephones

As in all communications the tendency is for the new systems to be digital. The second generation of cordless phones was, therefore, no exception and the standard produced for these systems is known as CT2. The specification can be used not only for office and home telephones but also for the public network known as telepoint. In many countries base stations are being installed in streets and places such as railway stations and shopping centres which allow a subscriber with the correct type of phone to make calls from within 200 metres of a base station. In addition to CT2, ETSI is producing another digital standard known as DECT (Digital European Cordless Telephone). This is being developed for office use where traffic is more intense. However, the same specification can be used for telepoint and home use. The DECT standard is very comprehensive and has been prepared to support all applications where cordless telephones may be involved.

The cordless telephone in an office or factory is obviously attractive as a fixed installation is no longer required and calls to extensions can be directed to the person rather than to an empty desk.

The development of the CT2 digital standard in the UK was sponsored by the Department of Trade and Industry and the work was performed by a number of companies. The standard was prepared in two parts. The first defined the speech transmission and signalling between the base station and the handset as if they were directly connected and the second part defined the radio parameters between the units. With the introduction of telepoint and the need for compatibility for all handsets when roaming it became essential to define the Common Air Interface (CAI). This CT2/CAI specification, MPT 1375, was made in 1989 and is available free from the Radiocommunications Agency. It is a large book and obviously only brief details can be included in this book. The DECT specification is even larger.

The dispute in 1987 which prevented an agreement in Europe on a common digital standard arose over the the use of TDMA (Time Division Multiple Access) and FDMA (Frequency Division Multiple Access) and eventually it was decided to produce the DECT specification under the control of ETSI which would operate at approximately 1.8 GHz using TDMA. CT2/CAI using FDMA also became an approved European standard.

There are many common features with CT2 and DECT. These include the following:

Both are digital for the operation of cordless telephones.
Both use time division duplex (TDD)
Both use the same speech coding (CCITT G.721 32 kbits/s ADPCM speech coding algorithm).
Both have a similar RF power output of 10 mW.

The differences are, however, in the frame structures and facilities.

CT2

This uses 40 separate carriers to provide 40 duplex channels. Each pair of send and receive channels is multiplexed onto a carrier in time division mode (FDMA/TDD). Time division duplex creates the illusion of continuous duplex transmission and signalling although only one channel is used alternately in both directions.

CT2/CAI is specified to operate in the band 864.1 to 868.1 MHz. The signal is transmitted as a two-level FSK waveform (see Chapter 12).

One major advantage that the digital system has over the analogue systems is that there is privacy for the users. It is difficult for another telephone to overhear casually an intelligible conversation.

The CT2 system has been developed to provide a cheap high quality digital voice service for many applications. It is relatively easy to engineer into systems and the specification has been written to allow additional services to be added by manufacturers. CT2 is now an established system and equipment is readily available. Telepoint services are being established using CT2.

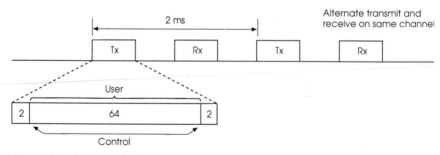

Figure 13.1   *CT2 frame structure.*

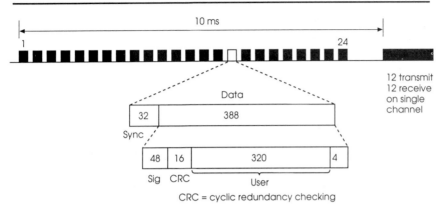

Figure 13.2 *DECT frame structure.*

DECT

DECT is a combination of all three methods: FDMA, TDMA and TDD. It uses 10 separate carriers to provide 120 duplex channels; 12 send and 12 return channels are multiplexed onto each carrier. The channel spacing is 1728 kHz and it uses a bit rate of 1152 kbits/s.

From the outset DECT was specified to provide a wide area of applications in both large and small business communications, telepoint services and residential applications. Provision is made for both voice and data and future ISDN services should be accommodated. These include circuit switched voice and data and packet switched data. A DECT network consists of both portable and fixed parts of a radio system connected by an air interface. The system should be capable of directly connecting to a global network such as GSM and telepoint or a local network. The DECT network can even be part of a cordless PABX or other similar equipment.

The DECT network supports a number of communication channels known as bearers. The basic bearer is 32 kbits/s but time slots can be linked together to produce high bearers of higher capacity in either simplex or duplex: 12 time slots can produce a bearer in simplex mode of 384 kbits/s. This would be necessary for ISDN and LAN applications (Chapter 7). It is also possible for lower rates than 32 kbits/s to use half time slots. These are required for the low bit rate codecs being used for speech. Use is made of the ISO model to define the different parameters of the standard.

The transmitted bit rate of DECT is higher than CT2 making the bit period shorter. This period is equivalent to a propagation delay of 345 m and this may require equalisation in outdoor applications.

DECT is far more comprehensive than CT2 but also more complex and requires the support of more software. However, by an EC Directive each country must allocate part of the spectrum for DECT systems. This does not apply to CT2. DECT is at present far behind CT2 in the development of equipment and systems for telepoint and offices and CT2 will remain the cheaper option for equipment not requiring the exacting specification of DECT.

Common Air Interface (CAI)

Several companies were producing CT2 equipment by 1988 but there was a variety of air interfaces which had been developed to satisfy the different environments in which the equipment was being used. With the introduction of telepoint it was necessary to standardise the interface to allow different handsets to roam on the systems.

There are a maximum of 40 channels in the system. Each channel has a bandwidth of 100 kHz and on this is modulated the encoded speech of 32 kbits/s in each direction together with signalling of 1–2 kbits/s. A guardtime for transmit and receive switching must be provided in both directions. Up to 1 bit must also be allowed for the propagation delay for remote antennae or a leaky feeder. A compromise specification provides a frame period of 2 ms and an instantaneous data rate of 72 kbits/s. This means that each base station and handset transmits for 1 ms in turn to each other at a data rate of 72 kbits/s.

A time division duplex system requires both bit and frame synchronisation between the base station and the handset. This has to take place during call set-up and the CAI specification combines this with call indication.

When transmitting data the base station must be transmitting when the handset is receiving and the process reverses for the other direction of transmission. The two transmissions occur every 20 ms and are controlled and synchronised by the base station. The base station is chosen as all calls into the base station must be synchronised to the same pattern in order to avoid mutual interference. Synchronisation is achieved by the base station transmitting every 2 ms and the handset locking to this pattern.

Both the base station and handset transmit 66 bits (or 68 bits if capable) every 2 ms at a rate of 72 kbits/s. Each burst is shaped at the start and finish so that the amplitude builds up and decays. This is to prevent modulation products being produced due to rapid switching.

Power control of handsets is essential in order to ensure that a handset close to a base station does not desensitise a base station and prevent communication with other users. All handsets must, therefore, be capable of operating with two power levels. Normal power is between 1 and 10 mW and low power is 12 to 20 dB lower. Base station commands decide if a reduction of power is necessary.

Telepoint - Rabbit system

Although several companies were licensed for a telepoint service in the UK only one continued with the installation of the numerous base stations required for such a system. Unfortunately the system was not a commercial success in the UK and ceased in December 1993. It does, however, continue in other countries. The concept of telepoint is that a small portable phone can be used in the home and office with a small base station and from within 100 metres of fixed base stations situated in streets, railway stations, airports and all other places where sufficient people congregate. The transmission system between the handset and the telepoint can be designed to allow users to register their presence in an area and receive calls but this approach was not attempted in the UK. Calls using telepoints in the UK were only originated by the subscriber. The system, however, can be combined with a paging system which informs a subscriber of a contact number which needs a reply. The subscriber can then decide when and how to contact the caller.

The commercial success of telepoint depends on educating the public about its advantages. The ordinary person is probably happy with a home and office phone and the use of a call box when making a call away from home. The business person now has the choice of cellular networks for TACS, GSM and PCN, as well as extensive paging systems and data networks. With such a choice of systems and assuming there are no major engineering problems that cannot be solved, the success of individual systems will depend upon marketing creating a need and the cost to the subscriber of using the individual systems.

Hutchison, who installed the Rabbit system, chose CT2 and the Common Air Interface (CAI). Since the publication of MPT 1375 the standard has been adopted in at least 15 countries and receives the manufacturing support of many major companies such as Panasonic, Northern Telecom, Sony, Motorola and Siemens amongst others. Hutchison also launched the equivalent system in Hong Kong, where it is called Tien Dey Seen, with apparent success. Telepoint systems are receiving trials or being installed in several countries throughout the world, including France, Singapore, Finland, Germany, Australia, Canada, China, Thailand and many more are planned. There is no standard for how the different systems operate and the facilities they offer although the equipment operates to the CT2/CAI specification.

The Rabbit network

Although no longer operating in the UK, similar systems operate in France, Hong Kong and Holland. The Rabbit phone was sold as a base station for the home and office with the facility of using the handset with its own base station or one of the telepoints installed around the UK. In order to use the telepoints the user had to subscribe to the system. The complete system was designed to reduce costs to a minimum and facilities were tailored to this end.

The UK system did not allow a subscriber to move between telepoints and maintain a continuous conversation. The subscriber had to remain with their originating base station. The French system is the same but in airports and railway stations a required handover between local base stations can be detected and the call can continue uninterrupted when the subscriber is moving.

In order for maximum sales to the public the Rabbit system was sold from high street shops as a cordless phone with high quality speech and the added telepoint feature if required.

When the phone was purchased there were three ways in which the user could be confirmed as a subscriber to the Rabbit system. All of them require the system to have a large database which could be accessed to confirm if the user was registered with the network. A subscriber's details were registered on the network computer when a subscription was paid.

Base station antenna ⟶

An omni-directional antenna was installed over the Barclays' sign and provided coverage in several directions. In tube stations directional antennae were used which allowed the system to be used underground.

Figure 13.3   *A typical base station for the Rabbit system.*

When the phone was first purchased a code locked the possibility of dialling and it was necessary to obtain a PIN number from Hutchison to unlock the phone. When the caller made the first call on a telepoint the registration information previously programmed into the handset was verified and then modified via the base station for future use.

A future specification would have avoided the necessity of programming each phone before sale and when the first call was made the registration information could have been sent from the computer and loaded into the handset.

The alternative to registration information in the handset is the necessity to enter manually a long code from the keyboard. This occurred when foreign units from other systems used the UK telepoints.

The base stations were connected to the PSTN and data was transmitted on a data network. The number of channels operating was a maximum of 40 on any base station but such a number would rarely occur. The base stations were provided in units of two channels and additional channels were provided if traffic demands increased.

CT2/CAI has the immediate advantage over DECT that systems are already installed and working. Although DECT would provide the same facilities it can be more complex than is required for many systems and more expensive. It will be the cost paid by the subscriber that will eventually decide the long term success of any system if there are alternative ways of providing the service.

Index